Information Analysis, Consolidation and Repackaging

Sewa Singh

ATLANTIC

PUBLISHERS & DISTRIBUTORS (P) LTD

Published by
ATLANTIC
PUBLISHERS & DISTRIBUTORS (P) LTD
7/22, Ansari Road, Darya Ganj,
New Delhi-110002
Phones : +91-11-40775252, 23273880, 23275880, 23280451
Fax : +91-11-23285873
Web : www.atlanticbooks.com
E-mail : orders@atlanticbooks.com

Branch Office
5, Nallathambi Street, Wallajah Road,
Chennai-600002
Phones : +91-44-64611085, 32413319
E-mail : chennai@atlanticbooks.com

ISBN: 978-81-269-1861-4

Printed in India at Glorious Printers, A-13, D.S.I.D.C.,
Jhilmil Industrial Area, Delhi-110095

To the memory of
my parents

Preface

In this age of Information Technology, information has assumed greater importance in not only everyday human activities but Research and Development also. It has become a vital resource and valuable input in socio-economic development of every country because it is collecting, organizing, and disseminating information for its users. However, the tremendous growth of information has, inter alia, become a barrier to its effective use. This has created problems both for information users as well as the information providers and specialists. To cope with such a situation the librarians and information professionals developed such processes in bibliographic organization as classification, indexing, abstracting, etc. but these have been found inadequate and ineffective in making information accessible in the emerging electronic environment. The recent developments in information science leading to the emergence of new methods and processes, helping in designing more effective information services and products, were found to be more helpful in the fruitful use of information. The concept of information consolidation and repackaging was developed in conjunction with the concept of appropriate technology and technology transfer to provide relevant and concise information to the end users in different subjects.

Information consolidation activities may be taken as responsibilities exercised by individuals, organizations, etc. for the process of evaluating and compressing the requisite information for ultimately providing to various user groups in usable form.

This book is an outcome of my lectures delivered to the postgraduate students of Library and Information Science in various universities over a period of years. The subject matter has been organized in 15 chapters to cover all related aspects of the process of information consolidation and repackaging. It provides an overview of information and information consolidation giving an idea about the basic concepts. The methodology of preparing information products by librarians/subject specialists in an organization has been explained. The book further describes the steps to be followed in the preparation of various information consolidation products such as handbooks and manuals, newsletters, state-of-the-art reports, reviews, trend reports and technical digests. It also provides an overview of the planning and management of an information consolidation unit as well as the knowledge and skills required for information analysis and consolidation. Towards the end are provided criteria and steps for evaluation of information consolidation products as well as some of the emerging trends in information analysis, consolidation and repackaging.

A sincere effort has been made to make the book useful for the students, teachers and working professionals. For any gaps in it, however, the responsibility is entirely mine.

I would like to acknowledge the help of the authors whose books and articles I have used freely in this book. I am particularly grateful to Dr Nagamohan Rao, Scientist, DRDO, Hyderabad, and to all other friends, colleagues, and students who have helped me in one way or the other. I thank Dr. K.R. Gupta, Honorary Adviser, Atlantic Publishers and Distributors (P) Limited and Mr. Harjeet Singh, Publishing Co-ordinator for publishing the book in a short time.

Amritsar **Sewa Singh**

Contents

List of Figures and Tables

Figures

Tables

Information Analysis

1

1. Concept of Information

In modern society, information has been recognized as a vital source of social and economic significance. It has always been considered as an important component of social change, economic growth and development, but in recent times information has assumed greater importance. Development is a purposeful social change toward a kind of social and economic system that a society wants. It involves introduction of new ideas particularly concerning modern methods of production of economic goods, new practices in such sectors as health, education, etc. so as to raise the quality of life in the society. Development is therefore not only a complex but also evolutionary process that needs positive interaction of a multitude of elements requiring such resources as human, economic, physical, technical, and informational, among others.

Information, particularly scientific, technical, commercial, health, education, etc. is a vital resource needed for socio-economic development. Information is the creation of human brain in action and it has great potential value in making right decisions and problem solving, at the governmental, institutional, and personal levels. Thus, the use of information is the main factor in making it a valuable resource; and if it is not used it is as useless as buried, lost treasure. The use of information, in turn, depends to a considerable extent on the availability and accessibility of right information and its proper explanation which gives a society and a nation the social, political, economic, and technological advantage, especially in developing countries.

Efforts have been made, particularly since the second half of the twentieth century, to develop sufficient information infrastructure in order to improve the availability and accessibility of information in most of the developing countries. These efforts have succeeded, to some extent, in many of these countries, though much needs still to be accomplished as the utilization of information has not reached its potential as yet.

We need to take a bird's view of the concept of information, and appropriate information, lack of which perhaps becomes a problem in the way of utilization.

2. Information: Definitions

Information is a phenomenon as basic as energy and matter. It is considered as a processed data as well as a representation of facts or ideas. According to J.H. Shera, information, both in the sense it is used by the biologists and in the sense librarians use it, is a 'fact'. It may be an isolated fact or whole cluster of facts; but it is still a unit of thought. This is what G. Bhattacharyya stated that information is a message conveyed or intended to be conveyed by a systematized body of ideas.

On the basis of Claude E. Shanon and Warren Weaver's information theory, information is defined as "a selection from a set of available messages, a selection which reduces uncertainty. Information is that which relieves uncertainty." Accordingly, information is an invariant property of a message, a signal, or a document.

However, information has been variously defined, but it is observed that a comprehensive definition is not possible due to its amorphous, complex, and multifarious nature. The nature of information is that it is an element of knowledge spectrum.

But everyone seems to deal with information in many ways—the whole life. In this context D.A. Kemp has stated that "Indeed information has been described as the fifth need of man ranking after air, water, food, and shelter."

However, some of the definitions are given to have a broad understanding of the concept of information.

According to J.H. Shera, "Information is that which is transmitted by the act or process of communication. It may be a message, a signal, a stimulus...."

Harrod's Librarians' Glossary and Reference Book has defined information as "an assemblage of data in a comprehensive form, recorded on paper or some other medium, and capable of communication."

ALA World Encyclopedia of Library and Information Service states that "Information is a property of data resulting from or produced by a process that produced the data."

The above-mentioned definitions make it amply clear that information collection, transfer and use are all-pervasive and universal activities in all walks of life. All of us, most of the time, are concerned with information, i.e., its generation, recording, processing, repackaging, transfer, receiving, use and application. In the present times, creation of new knowledge and information, their processing, storage, retrieval, dissemination, distribution, etc. have become critical areas for industrial investments; indeed today information industries are emerging as a major group among modern industries. According to an estimate, in the USA the Research and Development organizations invest about $ 30-40 million in a 30-year career of a scientist or engineer.

Probably, this is why Anthony Debons opined that "information enfolds man physically and spiritually. It is the vehicle and substance...(and)...is a resource...for...power." But information becomes power only when it is communicated or transferred to the user. In other words, information by itself has no inherent value. It has value only when used in some decision process.

3. Barriers to Information

The application and use of information has been given highest priority in recent times, particularly among the Western industrialized countries. The developing countries, on the contrary suffer from a number of constraints in the acquisition, storage, processing, and making information available for use. Even the transfer of information from developed to developing countries is not as smooth as there are a number of hindrances in the free flow of information. These barriers to the access of information are likely to prevent communication and free flow of information affecting individuals, groups, as well as organizations.

Some of the barriers in the free flow and use of information, in such areas as science, technology, health, business, etc., are discussed below.

3.1 Language Barriers

There are innumerable languages spoken in the world to facilitate transfer and use of information. But it becomes a barrier in the flow of information as the scientists, academicians and other people know only a couple of languages, and can communicate with limited peer groups. Though language is a powerful means of communication, but when the information is presented in a language and terminology which is beyond the comprehension of the users, like the problems of jargon (such as neologism, synonym, acronym) as well as that of presentation (of ideas, its level, style, form, etc.) do not help in communicating the intended message to the receiver.

3.2 Cultural and Social Differentiation

Cultural and social differentiation may cause serious problems of information communication and transfer. These differences could be with regard to level of development of different countries, their cultural concerns, differences in ideologies, and so on. Problems with regard to communication of ideas could also stem from multiplicity of sources, seepage and scatter of information. This problem is further complicated by the media comprehension, its perception, alien of reality, misunderstanding of the information to be communicated, etc.

3.3 Technical Barriers

Some of the major technical hindrances in the free flow of information include: lack of special equipments; poor presentation of documentary products; lack of proper organizational structure; under-qualified personnel; lack of awareness about information services among users; complicated systems; and so on.

3.4 Financial Barriers

The cost of production of information, its acquisition in library and information centres, its processing and distribution, etc. goes up because of financial barriers coming its way. These barriers could be in the form of depleting budgets, increasing cost of production, currency exchange rates, increasing cost of hiring library personnel, buying books and other reading materials, equipments, etc. contribute to rising costs of scientific information.

3.5 Psychological Barriers

Library users generally face many psychological problems in accessing the right information at the right time. They may be reluctant to ask for the required information from the library and information professionals for reasons of shyness, or ignorance, or mistrust, or false prestige, or unwillingness to accept the changes, and so on.

3.6 Institutional Barriers

Free flow of information is also sometimes impeded by the institutional hindrances. In various types of organizations when it comes to collection or gathering of information it generally comes from persons lower in the ranks and hierarchy and is communicated to the top management. Such barriers are not considered helpful in the use of information for the benefit of the institutions or organizations.

3.7 Administrative Barriers

A variety of administrative problems come in the way of flow of information in different institutions. These difficulties may be associated with regard to the policies, procedures or other types of regulations of the concerned institutions.

3.8 Packaging Barrier

Sometimes information is presented in a form that is not easy to follow and comprehend. The packaging of information may be an impediment to information absorption and purposeful use.

3.9 Validity of Information

On many occasions, information may not be trusted by the potential users. It may be because validity and reliability of information is not evaluated and thus information is questioned by the actual and potential users.

3.10 Information Overload

In the present times, there is too much of information available on a topic and the potential user is overloaded by its presence. The sheer amount of information decreases the willingness to use information, taking too much of time and effort on the part of the users.

4. Solutions to Information Barriers

These and similar other barriers have been and still are the most serious impediments to the use of information throughout the world. Many solutions have been proposed to remove these impediments, and some of them have also been implemented in certain information systems. Some of the solutions lie in various secondary information services as well as bibliographic organization, classification, indexing, abstracting, etc. However, few other solutions to this problem, as suggested by Tefco Saracevic, have been discussed below.

1. **Scientific and technical reviews,** state-of-the-art, and trend reports are considered quite useful in removing the barriers in the flow of adequate information.
2. **Handbooks:** A variety of handbooks synthesizing data from a number of sources are helpful in removing the barriers.
3. **Compilations:** Various types of compilations of critical, evaluated data are also of use in this regard.

4. **Summaries:** Business, commerce and market summaries help greatly in the flow of required information.
5. **Briefings:** Briefings on a topic requiring a decision and discussing alternate courses of action are considered useful.
6. **Technical writings:** If technical writings producing information on a complex technical topic are produced in simplified language they could be used widely by non-specialists as well.
7. **Trend reports:** These are generally based on a review of documents on current developments in a subject, hence useful for subject specialists.
8. **Technical digests:** These are considered useful sources of information for managerial and technical workers in an industry.
9. **Popularization of science writings:** For the purpose of popularization of science writings producing information on scientific advances and cause-effect findings, these must be made comprehensible to non-specialist users.
10. **Dissemination of information:** For free flow of information its dissemination through demonstrations and show-how is required.

The information services and products as enumerated above are designed to afford a fruitful use of appropriate information by the users. Some of these products and services are addressing the needs of specialists, like scientists, engineers, doctors, etc., while others are directed toward broader groups in the population, like workers, farmers, technicians, etc.

Library and information centres must take initiatives to establish information analysis units so that suitable information programme could be developed in order to provide evaluation and synthesis of existing information products as well as for designing new ones to meet the rising information needs.

It is from the products and services as enumerated above that the concept of information analysis and consolidation has evolved. It is also experienced that there is a definite need for information analysis and consolidation of products and services in the present as well as emerging information environment.

5. Information Analysis and Consolidation

5.1 Concept, Need and Purpose

The activities covered under the umbrella term Information Analysis and Consolidation are not new to the subject specialists. They have been making use of these activities and the resulting products like handbooks, data compilations, digests, etc.

In its final report, UNESCO *Symposium on Information Analysis and Consolidation*, held in Colombo (1978), made clear the definitions of the terms "analysis" and "information consolidation unit" in relation to their particular function. As per the observations made in this *Symposium*, the term "analysis" comprised a wide range of functions, such as abstracting, indexing, translation, reviewing, consolidation, etc. But because many analysis centres do not always perform the consolidation function, and information consolidation activities can also be performed within institutions or systems other than information analysis centres, therefore the term "information consolidation activities" should be used to define:

> ...the responsibility exercised by individuals, departments or organizations for evaluating and compressing relevant documents in order to provide definite user groups with reliable and concise new bodies of knowledge. Individuals or groups of individuals performing information consolidation activities would each constitute Information Consolidation Unit (ICU).

This definition makes it clear that evaluation, i.e. the retaining of reliable information, is the most important function of information consolidation. Information consolidation is therefore, different from traditional forms of intellectual organization of information which includes indexing, abstracting, classifying, cataloguing, etc.

Tefco Saracevic and Judith B. Wood have provided a more comprehensive definition of consolidated information as under:

> Consolidated information is public knowledge specifically selected, analyzed, evaluated, and possibly restructured and repackaged for the purpose of serving some of the immediate decisions, problems, and information needs of a defined clientele or social group, who otherwise may not be able to effectively and efficiently access and use this knowledge as available in the great amounts of documents or in its original form. The criteria for selection, evaluation, restructuring, and repackaging of this knowledge are derived from the potential clientele.

They have further given a little more concise and generalized view:

> Consolidated information is a text(s) or message(s) purposefully structured from existing public knowledge to affect the private knowledge and decisions of individuals who otherwise may not be able to effectively and efficiently access or use this public knowledge from the original amounts or in the original structure and form.

The concept of information analysis is not the result of new developments. The roots of information analysis and consolidation activities have been embedded deep in the efforts of eighteenth and nineteenth century European scientists. These scientists, who included German mathematician Gottfried W. Leibniz, Swedish and German chemists Jons J. Berzelius, and Leopold Gmelin respectively, and Konrad F. Belistein, critically evaluated significant scholarly work of the time, synthesized data and literature, and published for the benefit of the fraternity. Indeed, each of these scientists was a one-man information analysis centre or information consolidation unit who seem to have contributed to their fields by information evaluation and consolidation.

There are three basic contemporary problems concerning information, i.e.:

(i) **Scatter:** Information on a particular subject or information which may be useful in solving particular problems is often scattered all over the world in a variety of information sources.

(ii) **Amount:** The amount of information is growing at a rapid pace in a great many sources due to proverbial information explosion.

(iii) **Quality:** However, many of these information sources are of uneven quality, while some of them are redundant, and some others are of questionable quality.

In order to cope with these problems a number of systems have been instituted and many kinds of efforts undertaken at individual and organizational levels. One such effort is information analysis and synthesis as performed by information consolidation units.

Saracevic and Wood have given the definitions of analysis and synthesis as follows:

> Analysis is a process of determining and isolating the most salient information conveyed by a given information source and separating this information into its constituent elements on the basis of predetermined evaluative and other criteria.

They have further observed that:

> Synthesis is a process of condensation and distillation of analyzed information from one or more sources and presentation of information in a new arrangement or structure with an interpretative or evaluative point of view.

Broadly speaking, analysis is involved in any and all information representations, i.e., indexing, abstracting, cataloguing, etc. On the other hand, it is often seen that users looking for information are unable to identify specific information required by them because their information needs are not yet fully defined. It is also possible that the users may not understand the retrieval system properly. They will therefore, not be able to translate their requirements into specific subject terms.

In the information representation activity, subject analysis and content analysis are two major steps which involve, as mentioned above, abstracting, indexing, etc.

Subject analysis is usually performed to meet the information needs of a particular category of users who are using a particular information service or sources. But evaluative subject analysis involves not only deciding what a document is about but also why it is likely to be of interest to a particular group of users. In other words, there is not a set of correct index terms for any document, rather the same item could be indexed differently for a different category of users.

Content analysis is a research technique used by information scientists that helps to change the structure of the original document in order to get the required information in a manageable and acceptable format. Further, it should also be easily comprehendible by the user group for whom it is prepared. Thus, content analysis which is meant for retrieval purposes complements other social science methods of investigation.

6. Information Representation

Information representation is that aspect of information retrieval in which the original file of documents is represented by a set of tags or surrogates such as abstracts or index terms. The physical forms of representation are organized in such a manner that they may be manipulated and searched to access the contents more efficiently and effectively. The key concepts are organization of information resources in order that searching can be facilitated. The underlying aim of organization is for the express purpose of expediting information retrieval. Indexing and abstracting, therefore, are vital components in the communication link between the originator of information and its end user.

In order to extend help to specialists in their respective fields of interest (e.g. science and technology, social sciences, and humanities), abstracts and indexes are useful to identify the relevant documents.

7. Abstract

An abstract is an abbreviated, concise and accurate representation of the significant content of a document presenting its objectives, scope, and major finding, in a style similar to that of the original document. Besides, an abstract covers the main points made in the original document in a condensed manner usually following the arrangement of the original document.

According to Robert Collison, an abstract is "the terse presentation...of all the points made in the same order as in the original piece of primary documentary information...". It is usually accompanied by an adequate bibliographical description to enable the original document to be traced. Thus, an abstract is a document surrogate.

The primary objective of an abstract is to capture the essential content of the document thereby saving the time of the user. Instead of reading the entire document the user may decide about its relevance by reading the abstract which is a short representation of the document. In other words, an abstract helps the user in determining whether there is any need to consult the full text in order to gain further information about the subject.

8. Index

An index is a systematic guide to items contained in or concepts derived from a collection. It is basically a list of index entries used by information professionals as well as users for information retrieval. Each index entry leads to an indexed item outside the index, for example a record in a database, a journal article or a book in the library. The index entries are arranged in some recognizable order, usually alphabetical, and point to a particular document. In short, an index is an indicator of content and location.

The subject approach to document is solved by a two-step process. In the first step, each document is analyzed by an indexer according to the subject matter and assigned to one or more concept classes represented by index terms. In the second step, the appropriate index terms are chosen to represent the concepts identified in the document.

9. Subject Analysis

Subject analysis generally includes vocabulary structuring and subject indexing. Vocabulary structure refers to the constructing of tools such as classification systems, subject headings lists, and thesauri, designed to facilitate the organization and retrieval of information. Subject indexing, on the other hand, refers to the application of a vocabulary, which may be more or less structured, to indicate the documents.

Subject analysis is a form of indexing where subject headings and index terms reflect the concepts which are surrogates for the physical documents. However, the greatest challenge in document representation is in the creation and maintenance of subject access to documents.

9.1 Preliminary to Analysis

The processes on the basis of which analysis and synthesis are performed start much before any actual analysis takes place. These preliminary and fundamental processes include the following:

(i) Study of the *subject area* or mission in which information will be analyzed and synthesized.

(ii) Study of the potential *users and uses* for which the analysis and synthesis will be done.

(iii) *Organization and systematization* of the contents or characteristics of the subject or mission, i.e. a prior creation of a table of contents, classification, typology, or codification for analysis and synthesis.

(iv) Consolidation of *objectives, resources*, and *constraints* of the system or work within which analysis and synthesis are performed.

(v) Determination of *evaluative criteria* for use as the base for analysis and synthesis.

Saracevic and Wood observed that without specific guidelines from these five areas no meaningful and rational analysis can be performed. In other words, analysis and synthesis cannot be performed in a vacuum.

9.2 Importance of Subject Analysis

It is realized that the foundations of subject analysis may make the catalogues, indexes, classification systems more efficient tools in reference service as well as information retrieval.

Another advantage is creation of document representation and maintenance of subject access to documents.

Again, in case of indexing periodicals, when good quality subject analysis is applied while indexing, it becomes possible to use very broad subject areas. Such indexing periodicals/ services usually have excellent coverage of the subject concerned.

Further, good quality indexes serve as excellent current awareness tools as well as retrospective search tools.

Lastly, in preparation of in-house databases, their construction as well as text retrieval from them becomes quite easy and efficient from user viewpoint. This can also help in conducting online and interactive search of databases.

9.3 Indexing Process

Indexing a document is not just picking up and assigning terms or keywords but is the result of professional activity carried out by people with proper training and experience. The indexing process includes the following:

(i) **Familiarization:** Familiarization with the subject content of the document or a set of documents is the first step the indexer has to take. For a good quality index, the indexer must have a thorough appreciation and knowledge of the subject and the nature of the contribution it is intended to make to the advancement of knowledge in that subject. Besides, the particular document that is being indexed has to be read carefully so that the contents are understood properly. Special attention needs to be paid to the title, table of contents, abstract (if available), illustrations, section headings, conclusion, etc.

(ii) **Subject Analysis:** At the second stage, the indexer is ready for analysis of the document just ahead of selecting index terms. The correct bibliographical data about the document according to a predetermined format will be recorded. The analysis of the document and decision regarding the subjects which are deemed to be significant involves tremendous intellectual activity. Therefore, only some features of the document, representing the main theme, can be specified, but for the rest the indexer has to rely, to a large extent, upon the experience, intellect, and may be intuition as well. Sometimes, some guidelines are provided to the indexer for the sake of uniformity and standard, but at others, the indexer has to depend upon judgment reflecting the requirements of the users. While these decisions are being made, the indexer notes down the concepts, either using terms directly from the document or looks for subject concepts within the title, abstract, or the document text itself, or he/she can consult some standard list of subject headings.

(iii) **Selection and Extraction:** Selection and extraction of the most relevant or salient features, and information not needed is filtered out.

(iv) **Conversion of Concepts to Index Terms:** The appropriate index terms selected by the indexer have to match the concepts to be indexed. These may be free text, natural language, or may be chosen from some kind of controlled vocabulary, i.e. a classification system, or a thesaurus, or a subject headings list.

Seetharama has observed that the term "Information Analysis" in the context of various activities in the field of information does not connote and denote one and the same idea operationally. While in the context of indexing it is subject analysis, it is considered to mean critical evaluation when refers to an information analysis centre. On the other hand, in the context of information consolidation, it is however, used primarily to denote a non-evaluative information analysis for the purpose of consolidating the results of analysis into a textual presentation in which the isolated units of input undergo substantial structural changes. The product, thus, always is factual, non-critical or non-evaluative, but informative in nature. It is a firm, coherent, and unified body of information.

He further says that this approach is slightly different from the UNESCO and other definitions, wherein it is emphasized that the most important consolidation function was evaluation and it is this function that distinguished information consolidation from processes such as indexing, abstracting, etc.

References

Atherton, Pauline A. *Handbook for Information Systems and Services*. Paris: UNESCO, 1977.

Saracevic, T. "Processes and Problems in Information Consolidation". *Information Processing & Management,* Vol. 22 (1) 1986: 45-60.

Saracevic, T. and Wood, J.B. *Consolidation of Information: A Handbook on Evaluation, Restructuring and Repackaging of Scientific and Technical Information*. Paris: UNESCO, 1981.

Seetharama, S. *Information Consolidation and Repackaging: Framework, Methodology, Planning*. New Delhi: Ess Ess Publications, 1997.

Shera, J.H. *The Foundations of Education for Librarianship*. New York: Becker and Hayes, 1972.

Weisman, Herman M. *Information Systems, Services and Centres*. New York: Becker and Hayes, 1972.

Information Consolidation, Packaging and Repackaging

2

1. Introduction

It is imperative upon the libraries and information centres not only to acquire and organize information but also develop techniques for its smooth transfer to the users. In this process of disseminating information to users, the way they want it, information professionals have to remove all the barriers coming in its way and make it as compatible with the users as possible. Alvin Toffler perhaps explained this dilemmatic situation in his famous book *The Third Wave*, when he stated that "...many people feel cut off because the very packages in which information arrives are unfamiliar...." Such situations have impressed upon the library and information professionals to make improvements in the easy and fast availability and accessibility of information to a large number of users in many developing countries as well. They have been developing adequate and appropriate information infrastructures, information systems, and application of information technology in library activities has further helped in achieving the goal of information accessibility.

Information needs and problems, however, are still great. In many cases the lack of information and the lack of resources for developing and up-keep of information infrastructure are still a challenge for the information professionals. Despite the good efforts, many a time, perhaps because of lack of appropriate information, the utilization does not reach the optimum levels. In such cases, appropriate information means the right information at the right time in the right amounts in the right format at the right level of comprehension by the different user groups.

Consolidation of information is, therefore, suggested as a solution to such problems caused by the lack of appropriate information in the right format and package. Information consolidation is regarded as a tool which aims to increase the utilization of information by the various user groups and help achieve the objectives of library and information centres.

2. Definition

A precise definition of information consolidation as suggested by Saracevic and Wood is as given below:

> Consolidated information is a text(s) message(s) purposefully structured from existing public knowledge to affect the private knowledge and decisions of individuals who otherwise may not be able to effectively access or use this public knowledge from the original amounts or in the original structure and form.

In other words, consolidated information is a particular form of representation of information where all the general aspects of information are still fully valid. There is a relation between the public knowledge and users having in between primary and secondary sources and consolidated information. Not all the primary and secondary sources of information may reach the users or be appropriate for users. Thus, the role of consolidated information becomes important in this context.

3. Essential Features

Saracevic has given the following features of information consolidation:

(i) User orientation, and

(ii) Cooperation between subject specialists and information specialists.

4. Objectives

Some of the basic objectives of information consolidation are summarized as under:

(i) To increase the effectiveness of information transfer.

(ii) To encourage more intensive information use for various developmental activities.

(iii) To widen the circle of potential users.

(iv) To fulfill specific information needs of users by providing evaluated and synthesized information.

5. Processes in Information Consolidation

Some of the basic processes involved in information consolidation are as given below:

5.1 Study of Potential Users

Users are the beginning and the end of all information systems. They are the key to the aims, direction and contents of any and all information consolidation activity. The information professionals have to identify the user groups not only before the commencement of information consolidation unit/activity but also during its life cycle. A user study is likely to increase the probability of a longer life cycle of such an activity.

Some of the steps in a user study include:

(i) Establishing need for user studies and their relation to all other processes.

(ii) Planning a user study.

(iii) Gathering information about characteristics of users.

(iv) Methodology of user study.

5.2 Selection of Information Sources

Selection of information sources from a variety of primary and secondary sources, potentially containing the most useful information for given user problems and information needs is a crucial process. The elements involved in this process include selectors and procedures.

Those given the responsibility of selecting information sources have to play a significant role as they make the ultimate judgments regarding selection of a particular source. In many information centres/organizations, selectors usually include subject experts, information specialists, librarians, committees involving users and information/subject experts, etc.

The tools for selecting information sources include bibliographies, selected lists, reviews, reference books, directories, indexing and abstracting periodicals, content tables, invisible colleges, websites, SDI services, etc.

The procedures meant for judging the value of information, procedures meant for judging utility of information sources, and use and appropriateness of the selected sources for definite user groups.

5.3 *Evaluation of Information*

According to Saracevic and Wood evaluation of information sources is concerned with the determination of its intrinsic merit, validity, and reliability. In other words, it means the quality of information sources which will eventually be consolidated. But to set up a criterion for evaluation of sources for information consolidation to cover all kinds of products for all categories of users is not an easy task because different circumstances demand different criteria. However, the following type of criteria may be used for evaluation purpose:

(i) Criteria in peer review of publications in science and technology includes analyzing objective criteria of evaluation. It includes analysis of a problem, hypotheses, concepts, methods, data analysis, interpretation, data validity, reliability, etc. Also to be studied are subjective criteria that include problem significance, solution significance, originality, reputation of the author/institution, value, reporting style, etc.

(ii) User evaluation of information services and products depends upon the criteria users themselves apply in judging the products and services received by them. The criteria most often applied by users include: information/data quality, scope of information, appropriateness, time lag, ease of usage/access, and costs involved.

5.4 *Analysis and Synthesis of Information*

Saracevic and Wood stated that "Analysis is a process of determining and isolating the most salient information conveyed by a given information source and separating this information into its constituent elements on the basis of predetermined evaluation and other criteria". In other words, once the information sources have been selected and acquired, they must be examined and further evaluated in order to extract the main points of information to be restructured and synthesized. It means that evaluative analysis and not only analysis to determine the contents as done in indexing and abstracting is significant.

Analysis of information requires the following:

(i) Specification of the topics in relation to which information will be analysed and extracted from a set of selected documents.

(ii) Specification of a scheme for organizing and systematizing information.

(iii) Details of procedures for extraction of the relevant information.

(iv) Assessment and verification of the extracted information into given headings or classes.

According to Saracevic and Wood, "Synthesis is a process of condensation and distillation of analyzed information from one or more sources and presentation of information in a new arrangement or structure with an interpretation or evaluative point of view".

From the above, it is clear that synthesis follows analysis, and it requires the following:

(i) Comparative arrangement of the extracted information from many sources.

(ii) Comparison of data from different sources, derivation of a consensus and resolution of possible conflicting information.

(iii) Compression or merger into a structure or form most suited for intended users and uses of information.

(iv) Evaluation of the final product.

Information is now considered ready for targeting and restructuring for use.

5.5 *Restructuring the Evaluated Information*

At this stage the results of user studies are available, information is selected and analyzed, and grounds for synthesis are prepared into a content or information product that can be used most effectively and efficiently by users to satisfy their needs. This may involve synthesis, condensation, rewriting, simplifying, review, state-of-the-art presentation, etc.

The information consolidation products are thus determined with one eye on the user and the other on the best available information on the given subject.

It requires the following decisions to be taken:

(i) Targeting to stages in information diffusion or transfer orienting specific products for specific users to stages and events in information transfer.

(ii) Restructuring the content aspects of synthesized information. It requires the determination of a mode and language in which the contents of information consolidation products will be presented to users in a way that is most compatible with users' situation and needs and that will enhance the comprehension and assimilation of information presented.

Restructuring of information requires the following:

(i) Extent to which available content is to be incorporated.

(ii) Degree of invariability of information as found in the original text, and degree to which new information is to be added.

(iii) Degree of details in information, i.e. specific to general.

(iv) Degree of changes in sequence of presentation relative to the original.

(v) Technical sophistication (i.e. minimal, moderate, high, very high).

(vi) Timeliness, time span covered, frequency of issuing the products.

(vii) Editorial qualities (i.e. grammar, clarity, balance, logic of presentation, etc.).

5.6 *Packaging and/or Repackaging*

The restructured information is to be packaged and/or repackaged in a form that will enhance the potential of its use. It may be pointed out that restructuring deals with the contents or substance of information, while repackaging deals with the media, format and the form of its presentation.

Saracevic and Wood define packaging of information as "a physical recording, arrangement and presentation of information on a given medium in a given form". In other words, it means that arrangement, shape and layout of information in a given product on a given medium. Selection of packaging aspects thus involves determination of the media and format in which an information consolidation product could be presented to users.

They further stated that the way information is packaged or repackaged cannot improve its content but it certainly can enhance its use, assimilation, and recall. The package in which information appears is often a decisive factor in acceptance or rejection of the whole, even before the contents have been considered; and it certainly is a factor in the amount of its use. The same content in different packages does receive different usage, does affect the rate of assimilation and does make recall dissimilar.

As a result of effect of technology, the basic media for packaging and repackaging include:

- Print media
- Audiovisual media
- Electronic media
- Interpersonal contacts (as a medium)

Each of these media has different requirements, criteria, procedures, economics, and technology for creation of information products.

5.7 *Dissemination of Information*

According to Saracevic and Wood: "Dissemination is a process (or processes) of conveying information through given channels to users; dissemination includes the spreading about, distribution and delivery of information products and/or services." There are a great number of ways and means by which dissemination can be affected. Some of the important channels, they suggested, for dissemination of information consolidated products include the following:

(i) Interpersonal delivery—products personally delivered to each user.

(ii) Group personal delivery—products given to users at meetings, demonstration, etc.

(iii) Strategic placement—products put at key sites for users to pick up, etc.

(iv) Local depositories—products circulated from a library, extension centres, etc.

(v) In-house dissemination.

(vi) Mass media—print.

(vii) Broadcasting—radio, TV.

(viii) Mail and other means of public distribution.

(ix) Telecommunications—telephone, satellite, etc.

(x) Computer networks—online systems, computer conferencing, etc.

It is worth mentioning here that the work of the Information Consolidation Unit is not complete with the developing of a product or provision of a service only. Choice of an effective channel of dissemination ensures that the product does not remain only on the shelves, or a newly launched service remains unused.

Marketing of information is an aggregate of many activities directed at satisfying human information needs and wants through exchange processes. Marketing involves viewing the whole information service or product from the point of view of final result, i.e., from the use and user point of view. The central concept in marketing is user oriented rather than product oriented. It involves the following:

(i) Identifying intended group of users.

(ii) Finding out about their needs and wants.

(iii) Satisfying their needs with the right product or service, supported by the right communication about the products, their benefits and uses, and availability at the right time and location.

5.8 *Feedback*

A feedback mechanism should be set up so that there would be an interactive communication between the Information Consolidated Unit (ICU) and the targeted user groups. Feedback is required from users and, evaluation of the efforts should be carried out, and adjustments should be made accordingly. The information thus collected may be utilized to know about the lacunae and shortcomings existing in the information consolidation products and the ICU as a whole. On the basis of the feedback received, the ICU can initiate necessary corrective action.

Seetharama observed that in some of the dissemination channels like interpersonal delivery, groups personal delivery, telephone, etc., feedback is built-in so as to ensure effectiveness. Such channels would, therefore, necessarily be costly.

A summary of the processes (as described above), elements, and relations involved are summarized in Figure 2.1 for a quick overview.

6. Packaging and Repackaging

As mentioned above packaging of information is a physical recording, arrangement and presentation of information on a given medium and in a given form. It is the bundling of products and services to address specific needs. Saracevic and Wood stated that two aspects, as given below, can be distinguished in packaging:

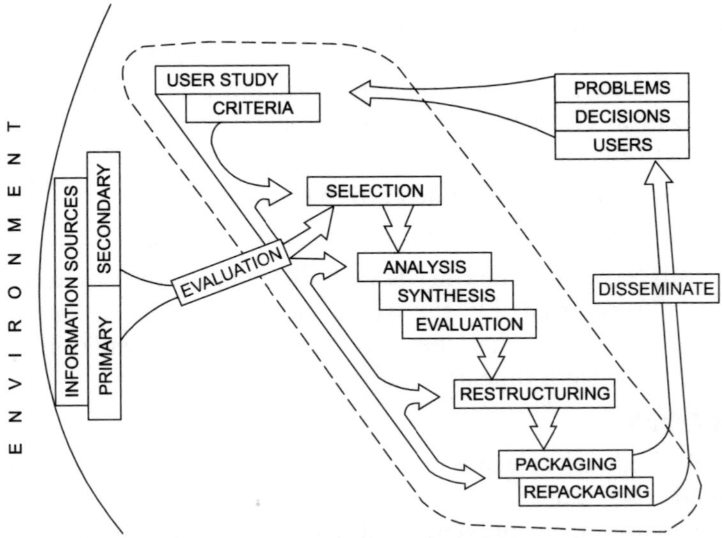

Fig. 2.1: Processes in Information Consolidation

Source: T. Saracevic and J.B. Wood.

(i) **Package Media:** The package media represents the physical medium in which the information is displayed or presented to users. For example, print, audio-visuals, demonstrations, interpersonal contact, etc.

(ii) **Package Formats:** The arrangement, shape, and lay-out of information in a given production on a given medium.

Packaging can be done by:

(a) Reformatting and synthesizing raw information.

(b) Combining expertise or consulting on a subject with access to relevant information sources.

(c) Providing training or assistance to a user in accessing an information product.

In order to add value to a product, the information provider must understand the types of information access problems most frequently encountered by the users. Based on this knowledge, packaging can add value or services that are not readily available elsewhere.

On the other hand, repackaging of information is rearrangement of physical media in which information has been presented, which is tailored to the requirements of a specific clientele. In other words, repackaging of information refers to the presentation of information in more understandable, readable, acceptable and usable forms.

The term information repackaging was first used by Saracevic and Wood, and Bunch in their publications in early 1980s. They described how for an information service appropriate material is selected, reprocessed to package and arrange information in a way that is appropriate to the users.

According to Greer, Agada and Grover, in the present information age, information overload can occur. Information repackaging can save time, labour, and costs to the user. It is a systematic process of adding value to information services. This is in line with the shift from document to their contents and from collections to their users.

Information repackaging can be done in many ways including reformatting and synthesizing raw information, combining expertise or consulting on subject with access to relevant information sources and providing training or assistance to users in increasing an information product.

6.1 Aim of Repackaging

The aim of repackaging is to enhance the acceptance and use of information products and the assimilation and recall of their contents.

6.2 Functions of Repackaging

The functions served by repackaging of information are:

(i) As a saving tool (time saver).

(ii) As a systematic and selective sorter of useful information.

(iii) As a means for more extensive information transmission and delivery.

(iv) As a translation tool.

(v) As an opportunity for the practical application of research results.

(vi) As a means for the prompt delivery of relevant information.

6.3 Importance of Repackaging

The importance of repackaging of information is reflected through the following points:

(i) Information repackaging entails a systematic process of adding value to information services.

(ii) These value added components would include but are not limited to information analysis, synthesis, editing, translating and transmitting its symbolic and media formats.

(iii) It ensures currency, accuracy, pertinence, comprehensiveness, ease of comprehension, and convenience of use.

6.4 Reasons of Repackaging

Some of the reasons of repackaging information can be identified as given below:

(i) To customize information to user needs.

(ii) To facilitate dissemination, organization, and for communication.

(iii) To simplify, i.e., an annotated bibliography is like a map in the world of information overload.

(iv) To facilitate interactivity between user, knowledge base, and technology.

6.5 Requirements for Repackaging

Sturges and Chimsen have listed three requirements for repackaging information:

(i) The materials should be collected and organized efficiently.

(ii) There should be the capacity to analyze their content and create new information package from them.

(iii) The new products should be disseminated freely.

This requires the identification of good models for repackaging which requires critical thinking for combining information from different sources, considering the accuracy, completeness, and consistency of the information.

7. Information Consolidation: Justification

In order to deal with the problem of increasing flow of data and information the scientists in the 18th and 19th centuries began to analyze, evaluate and synthesize the relevant information. This situation arose perhaps for the reason that the libraries and information retrieval systems of the time could not handle well the information in need. The traditional libraries, as Kertesz believed, were probably not geared for the tasks other than dealing with the documents and providing access to published literature. As a result, a need for information analysis and consolidation activities evolved wherein information contained in documents could be extracted and provided to the needy users.

The present day Information Consolidation Unit (ICU) is largely the result of the interest and efforts of numerous sponsoring agencies both in public as well as private sectors. Thus, a further need has arisen for justifying information consolidation activities to the satisfaction of the sponsoring agencies as well as the users.

According to Seetharama, justification can be discussed in relation to the following areas:

(i) Objectives of the parent institution;
(ii) Expressed or established user need;
(iii) Problem area;
(iv) Elaboration of the subject to be covered; and
(v) Proliferation of literature on the subject.

In order to understand their implications, these are discussed in the following paragraphs.

7.1 Objectives of Parent Institution

Most of the existing Information Analysis Centres and ICUs—whether they be discipline-oriented or mission-oriented—are, generally located in one parent institution or the other though they may be sponsored and supported by other organizations. Further, the activities of the ICUs seem to be in conformity with the objectives of the respective parent institutions. It may therefore be logically assumed that the institutions and their users having realized the importance and value of critical information for their research and other activities have established the respective ICUs. In other words, the scientific and/or technological, managerial and general objectives of the institutions concerned may have warranted the performance of information consolidation activities. For example, the technical management of a laboratory, or a sponsor, recognizes that in order to achieve the objectives of a new project, a certain kind of material must be available. This requires a large amount of evaluated information on different aspects which may not be readily available. Therefore, a need arises for the establishment of a well-staffed and organized ICU.

7.2 Expressed or Established User Need

On examining the features of the world of information carefully, the need for information consolidation activities would become obvious. Seetharama has presented the features of the world of information, the consequent user problems, and the information services or system capability required in a tabulated form as given under.

Table 2.1: World of Information

S/N	Features of the World of Information	Information users problems	Information service/system capacity
1.	Accelerated growth of information. Increasing rate of obsolescence	Inadequate time for reading and assimilation of information	Reviews, State-of-the-art reports, Trend reports, Digests, Information analysis and consolidation products.
2.	Wide variation in quality and realiability	Difficulties and inadequacy of time for evaluating and selecting	Data/Information analysis and evaluation
3.	Interdisciplinary nature of information. Scatter/ Seepage of Information	Can specialize only in a restricted subject field	Indexing techniques to interlink subjects
4.	Wide range of standards and modes of presentation of ideas	Only some standards and patterns are convenient to users	Selection and presentation or restructuring and repackaging according to user's needs
5.	Pertinent information published in documents with restricted circulation	Inaccessibility of documents	Location and procurement of inaccessible documents through channels other than formal

6.	Multiplicity of languages	Can be familiar with one or a few languages	Translation service
7.	Too many documents: Information Centres have limited funds; also space problem	Physical access to original document difficult	Reprography service

The above-mentioned problems, especially the overabundance of information, have made many a research scientist to create new information rather than search through existing documents, even though they may have time to do so. This has been mentioned primarily to focus the attention on the need for analyzed, consolidated and evaluated information to the users in their fields of interest.

The situation is the same in the non-R&D contexts, where the information user groups are the planners, policy makers, decision makers, managers, executives, industrialists, workers, small entrepreneurs, extension workers, rural people, etc. covering the whole spectrum from highly educated to illiterate levels. While managers require information for problem-solving and decision-making, the extension workers need information pertaining to the practical methods for spreading technology among its actual beneficiaries.

Recognizing the need for information, many information systems identify, collect, store, and disseminate information to the users. But this information may or may not be compatible with the users and their needs.

7.3 Problem Area

In some of the problem areas like Pollution, Biodeterioration, Nuclear Safety, etc. where existing knowledge and experience is scattered over several disciplines, and where the literature/information is spread over a wide variety of sources, it becomes difficult for researchers, managers, policy makers, to have access to information. In such problem areas, the interdisciplinary nature of information poses difficulties requiring information consolidation activities. The role of these activities is to seek solutions to complex problems, rather than to the general advancement of knowledge in an academic field.

7.4 Elaboration of Subjects to be Covered

Another reason for justifying the establishment and operation of an ICU is a sudden interest in a special field, such as engineering, or applied science, or space technology, for scientific or economic reasons. Such an elaboration of a subject requires special material for research and other academic activities. A well-staffed and organized ICU can be of great help to project management and researchers by bringing scattered references to their attention.

Similarly, study and detailed evaluation of newly developed methods or techniques can be best carried out in an ICU. Such an ICU would also facilitate correlation of all information in the field concerned. Thus, it amply justifies the setting up of an ICU for information consolidation activities.

7.5 Proliferation of Literature

The information situation today is described by such terms as information explosion, or information boom, or information flood, etc. Some of the factors which have influenced the character and increased the current pace of scientific and technological activities are:

(a) Exponential growth of publications;

(b) Rapid fragmentation, or Balkanization of knowledge;

(c) Interdisciplinary nature of scientific research;

(d) Increasingly effective technological capabilities; etc.

Such a rapid growth is characteristic of all subjects, especially the scientific and technological areas. This information explosion coupled with information "pollution" has created problems to users needing critically evaluated information, particularly in new, peripheral, and interdisciplinary fields. One of the methods of overcoming these problems is to establish an ICU whose task will be to organize, coordinate, and evaluate the information for its users.

It may, however, be mentioned that while information professionals are aware of the value of ICUs, the same cannot be said of the many users or potential users. Therefore, Saracevic and Wood have given a list of benefits which could accrue by the usage of consolidated information as justification of ICU.

Table 2.2: Potential Benefits of Consolidated Information

S/N	Activity or Area	Potential Benefits
	General Areas	
1.	Decisions	Better informed about alternatives and consequences. Improved decision-making process. Reduction of Uncertainty.
2.	Knowledge, Competence	Increase in level, depth, breadth by individuals or groups. Higher sophistication in drawing relations between seemingly unrelated facts.
3.	Adaptation	More appropriate and adjusted responses by individuals, groups or organizations to demands of and changes in the environment and a complex world.
4.	Productivity	Higher levels and outputs in work and other activities. More contacts. Increased capacity and effectiveness.
5.	Resources	More efficient or economic use of resources. Increased capacity and/or effectiveness. Better economy.
6.	Success	Contribution towards attaining aims of individuals, groups, organizations. Better or wider spread and acceptance of results. Detection of necessary adjustments.
	Specific Areas	
7.	Enterprises Business Commerce	Broadening markets. Providing appropriate responses to market conditions and demands. Meeting competition. Complying with standards and regulations. Motivating personnel. Making business decisions.
8.	Industry Manufacture	Incorporating technical innovation. Adding new products. Making products appropriate. Increasing productivity. Reducing breakdowns.
9.	Science	Keeping up with research front. Judging own position and advances. Making appropriate decisions on policy and allocations. Searching for related works and/or ideas for further work or methodology.
10.	Education	Keeping up with advances in given fields and in education, research, methods, approaches. Providing for educational planning and assessments or comparisons.
11.	Individuals	Providing for self-fulfillment and advancement in their given area of work or interest. Providing for wider opportunities in employment, self help, and adjustments to changing environment and conditions. Increasing sophistication towards higher quality of life.

8. Conditions for Transfer of Information

The basic aim of all information activities is for the utilization of information. According to Saracevic and Wood, the minimum conditions for effective transfer and use of information for socio-economic development are given as under:

(a) A propensity on the part of the decision makers and problem solvers to use information, which is based among other things, on recognition of the value of and need for information.

(b) A level of infrastructure (indigenous information systems and services) that makes the right information first available and then accessible for use.

(c) A type of political and social conditions which are favourable for encouraging the use of information and development of an information infrastructure.

However, these conditions are met by developing countries only partially and to highly variable degree. Therefore, they have synthesized a framework based on analysis of contemporary problems which is organized along the following aspects:

1. ***Recognition*** of the value and role of information in the process of socio-economic development.
2. ***Availability*** of information, particularly the definition of the kinds of information needed for development.
3. ***Accessibility*** of information, particularly the intellectual organization/representation of information.
4. ***Utilization*** of information and the requirements for increase in utilization.

Therefore, Saracevic and Wood believe that in the context of the above given framework, information consolidation can play a significant role. In this way, information professionals can make available and accessible to users information that is appropriate to the task at hand and packaged to user level, language, and available time. They view information consolidation as an intellectual post-processing of the available primary and secondary information to make it accessible to users, who otherwise would not access this information. Further, they recognize that information consolidation is just one in the spectrum of activities relevant to bring about an increase in information use for development. This is because of the fact that specific types of information consolidation are prepared to meet the specific user demands.

9. Value of Information Consolidation

Saracevic and Wood believe that the value of consolidated information lies in the fact that it can greatly affect social change in a positive manner. The aspects which influence the value of information are deeply imbedded in the social and cultural structure and greatly exceed the economic interpretations, as important as economic factors are. The value of consolidated information, therefore, is certainly economic. But the value may be even greater because it is related to other values in a society.

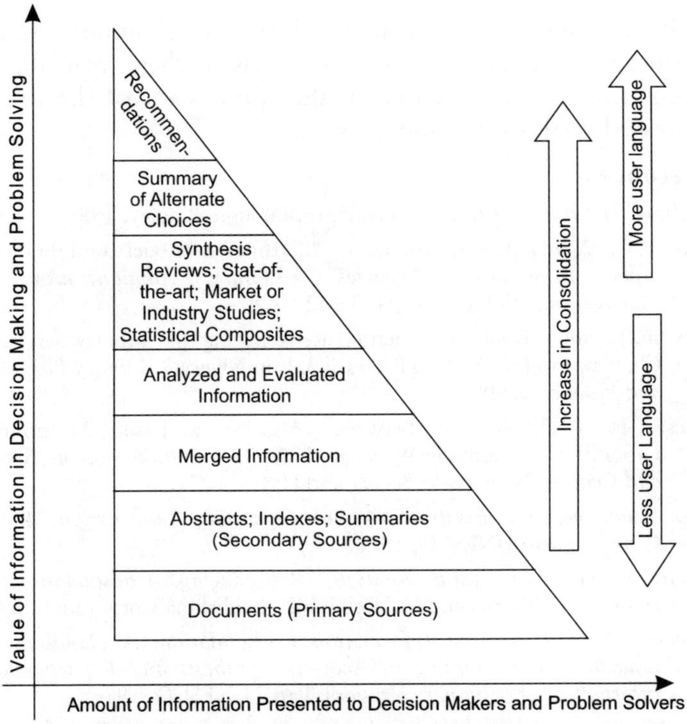

Figure 2.2: Value of Information in Decision Making and Problem Solving

The value of consolidated information is that it may contribute to information sophistication of a population, which in turn contributes to quality of life. They have argued effectively the value of consolidated information in relation to its role in decision-making and problem-solving. These relations are expressed in the above given Figure:

We can infer from the above that:

(i) As the amount of information presented to a decision maker is more and more consolidated, its value increases;

(ii) As the information is expressed more and more in the everyday language and social/cultural framework of the user, its value increases for that user;

(iii) As the information is more and more packaged in a way that will make its user easier, its value increases.

It can therefore be concluded that if communication is considered essential for social change, then consolidated information is also essential in the framework of the present status and goals of development.

References

Bunch, A. *The Basics of Information Work*. London: Bingley, 1984.

Greer, R.C., Agada, J. and Grover, R. "Staffing: A Model for Library and other Information Agencies". *Library Administration and Management*, Vol. 8(1) 1994: 35-42.

Iwhiwhu, Enemute Basil. "Information Repackaging and Library Services: A Challenge to Information Professionals in Nigeria". *Library Philosophy and Practice*, 2008.

Kertesz, F. "Collaboration Between IACs at a Large Multipurpose Laboratory". Quoted in Weisman, H.M. *Information Systems, Services and Centres*, New York: Becker and Hayes, 1972.

——. *Guidelines for Establishing and Operating Information Analysis Centres*. Paris: UNESCO, 1978.

Kumar, P.S.G. *Information Analysis, Repackaging, Consolidation and Information Retrieval*. New Delhi: B.R. Publishing Corporation, 2004.

Saracevic, T. *A Course in Information Consolidation: A Handbook for Education and Training in Analysis, Synthesis and Repackaging of Information (Preliminary Version)*. Paris: UNESCO, 1986.

——. "Processes and Problems in Information Consolidation". *Information Processing & Management*, Vol. 22 (1), 1986: 45-60.

Saracevic, T. and Wood, J.B. *Consolidation of Information: A Handbook on Evaluation, Restructuring and Repackaging of Scientific and Technical Information*. Paris: UNESCO, 1981.

Seetharama, S. *Guidelines for the Establishment of Information Consolidation Units*. Paris: UNESCO, 1984.

——. *Information Consolidation and Repackaging: Framework, Methodology, Planning*. New Delhi: Ess Ess Publications, 1997.

Sturges, P. and Chimsen, G. "Information Repackaging in Malawi". *African Journal of Library, Archives, and Information Science*, Vol. 6(2) 1996: 85-93.

Weisman, Herman M. *Information Systems, Services and Centres*. New York: Becker and Hayes, 1972.

Users and User Studies

3

1. Introduction

Users are an important component in an information system. They are the beginning and the end of all information systems. The information professionals usually take them for granted and at times they even remain neglected. Former US President John Kennedy once remarked: "The accumulation of knowledge is of little avail if it is not brought within the reach of those who can use it." But unfortunately, knowledge or information often is not brought within the reach of users. Therefore, information and information services that are not made available to the users are of no use. On the other hand, right information to the right user can pave the way for new directions to research and development. Information systems and services succeed if they adjust to users and their needs.

2. Users

It is imperative that to achieve this objective we should understand the library users, and how they interact with the system. Users are the last link or the end users, and the last recipient of the information in the communication cycle. According to Kenneth Whittaker a user may be defined as a person who uses one or more library services at least once a year.

2.1 Categories of Users

Users could be categorized according to the type of library, as the users of a public library will be different from those of an academic or research library. However, on the basis of their approach to information, the users can be categorized into four groups. According to Werner Kunz, et al. they are:

(i) **Potential Users:** They need information which can be provided by specific services.

(ii) **Expected Users:** They have the intention of using certain information services.

(iii) **Actual Users:** They have actually used one or more information services regardless of whether they derived advantage from it or not.

(iv) **Beneficiary Users:** They derive a measurable advantage from information services.

Similarly, there are other categories of users given by many authors including that of S.R. Ranganathan, who make use of various information services offered by the various information systems.

S. Seetharama has classified the users of information consolidation services into the following four groups:

(i) Researchers and Scientists

(ii) Policy Makers/Planners

(iii) Communicators/Intermediaries

(iv) Community Groups

Let us discuss about these user groups and their information requirements in the following paragraphs.

2.1.1 Researchers and Scientists

Scientists, researchers, engineers, etc. are some of the users of scientific and technological information. This group of users may be involved in research and development activities in an industrial setting, or they may be academicians engaged in teaching and research activities.

Pauline Atherton remarked that the work of these users must reflect the need for effective communication if it has to show the value of consolidated information. Therefore, their work must:

(i) stimulate thought and action by injection of, and interaction with, other people's ideas, knowledge, experience and achievements;

(ii) promote continuous awareness of what others are doing so that individual workers (or groups) may know of developments in their own special fields, and in wider fields, such as discipline, or technology;

(iii) diminish the probability of unwitting duplication of work and to save the time and effort;

(iv) provide specific information and data needed for work in hand; and

(v) meet specific requirements.

Voigt also expressed similar views when he said that "... scientists refer to information sources mainly in three circumstances" as mentioned below:

(i) while getting current awareness of results both in their particular narrow field and the related disciplines;

(ii) in their day to day work, when they need some factual information—figures, methods, and designs; and

(iii) when embarking on a new problem or a project, as well as when completing it and writing about it, a retrospective search to identify as many published and unpublished sources on the subject as possible.

On the other hand, Scherban had identified seven stages when information is supplied for scientific and technical work. These are given below:

(i) Overall familiarization with the problem, and problem statement. Drawing up a plan and the provisional terms for the solution of the problems of primary and secondary importance. This stage requires a general acquaintance with information available on the subject.

(ii) Gathering scientific knowledge about the subject of study. Retrospective searching of the broadest possible scope of literature without any pronounced critical approach.

(iii) Coordination and interpretation of scientific data. A critical evaluation of ideas and hypotheses of different authors. The relevance criteria for the information needed are specified and the amount of documents used is reduced.

(iv) Statement of hypotheses and choice of a working hypothesis, which is the most important stage of research in technical sciences. As to the use of information, it is characterized by in-depth analysis rather than broad coverage. Sometimes, however, a working hypothesis may be chosen under the influence of what might appear 'irrelevant' information.

(v) Proving the working hypothesis; the most important stage in basic research. The information used depends on the specific research; for instance, the proving of an assumption may require gathering factual data scattered in the literature.

(vi) Statement of conclusions and recommendations. Predictions, as well as generalizations, are frequently made at this stage. Information is often used to shed light on precedence and priority aspects.

(vii) Description of the research results. The information gathered and processed is, as a rule, minutely documented.

2.1.1.1 Types of Information Required

Further, researchers and scientists would require the following type of information:

(i) Scientific and technical information—enabling learning, research, technical decisions and actions.

(ii) Experimental information.

(iii) Know-why information.

(iv) Know-how information.

(v) Know-who information.

2.1.2 Policy Makers/Planners

Another important conglomeration of user groups is generally found in the non-Research and Development sectors.

These include planners, policy makers, decision-makers, managers, executives, industrialists, small entrepreneurs, extension workers, rural people, and so on covering the whole spectrum from highly educated to illiterate levels. These diverse categories of users also require diverse nature of information to meet their needs.

The information needs of some of these categories of users, such as planners, policy makers, managers, etc. would generally revolve round problem-solving and decision-making processes. Problem-solving concerns with the seeking of answer to a question, but decision-making is the cutting off of further consideration of the problem, the elimination of all alternatives but one and is essentially a commitment to action. Thus, decisions are made at a number of points in the sequential process of problem-solving. For example, decisions may pertain to:

(a) Goals which involve both values and empirical considerations;

(b) Sub-goals and means in means-ends chains of action;

(c) Ranges of input values;

(d) The selections from among alternate means-ends chains;

(e) Basic assumptions;

(f) Available data—should they be accepted or rejected, or should a search for further data be carried out; and

(g) Final action to implement the selected solution to the problem.

In other words, decision-making in its broadest context includes the following activities preceding the decision:

(i) finding occasions for making a decision;

(ii) finding possible courses of action; and

(iii) choosing among courses of action.

But, the essential ingredient for good and valid decisions should be the availability and use of information. More than sixty years ago, the importance of information in this context was well recognized by N. Weiner when he said that the recipe for good decisions is "90 per cent information and 10 per cent inspiration." Similar observation was made by G.A. Steiner when he remarked that "Information flows are as important to the life and health of business as the flow of blood to the life and death of an individual."

Therefore, to provide the planners, policy makers, managers, etc. relevant and appropriate information, ICUs should be set up and made operative so as to provide information in the following areas:

(i) engineering/design;

(ii) manufacturing/operations;

(iii) marketing;

(iv) finance/administration;

(v) project planning/control, etc.

2.1.3 Intermediaries/Communicators

It has been observed in the context of rural development that the ruralites, particularly in developing countries, are characterized by low per capita income, low rate of literacy, low rate of energy consumption, inadequate facilities for sanitation and transportation, etc. It has, therefore, been suggested to make use of appropriate technology as a dynamic approach towards rural development. For example, Biogas technology offers tremendous potential for rural development; therefore, conscious and sustained efforts have been made by the governmental and non-governmental agencies in developing countries for its promotion and adoption in rural areas. In this context need for extension activities and services has become imperative and relevant because self adoption of such technology, though found useful for rural people, has been difficult due to economic, social and cultural constraints.

Seetharama stated that the extension workers act as intermediaries functioning as development functionaries for establishing liaison between the generators/innovators and the actual beneficiaries. In other contexts, the other intermediaries may be teachers, mass media personnel, etc.

The objective of the intermediaries, whatever is their type, is to help people help themselves. Their focus is at changing the behaviour of the people, through voluntary participation, by changing their knowledge, skills, and attitudes for specific actions.

For the extension workers to function efficiently and effectively, they need detailed information pertaining to the technology; the practical methods for spreading the technology among its actual beneficiaries; the socio-cultural, geographical characteristics of the region concerned; the costs and benefits; etc.

Similarly, other intermediaries/communicators should be provided with appropriate information so as to make them functionally effective in order to achieve their goals. This necessitates the developing of an ICU for this purpose.

2.1.4 Community Groups

In developing countries, very little attention has been paid to the information needs of the majority who constitute the poorest of the poor people. They are the illiterate rural people who are estimated to be around 75 per cent of the Third World population. Information consolidation for such a section of the population will take very special unusual forms, as they cannot decipher the printed material.

J. Valls observed that little has been done so far to meet such requirements and it is about time that scientific and technological information contribute at least to the development process of the majority 'poorest' and not only to the minority 'elite'. As a matter of fact, many community groups are not aware of what the service agencies can do for them.

3. Information Needs of User Groups

From the above discussion, it may be concluded that the information needs of different user groups are quite diverse. It is therefore, incumbent on an ICU to cater to these diverse needs so as to fulfill its objectives. Saracevic and Wood have articulated the breakdown of information required for development needs. It is given below:

1. In terms of information types:
   (a) **Scientific and Technical Information:** enabling learning, research, technical decisions, and actions.
   (b) **Policy and Management Information:** enabling decisions on choices of alternate strategies and an increase in capabilities to manage, conduct business, negotiate, compare, evaluate, etc.; this may involve proprietary information as well.
   (c) **Operational and Industrial Information:** enabling production, maintenance, and services; this basically involves experimental information.
2. In terms of information use and transfer:
   (a) **Know-why Information:** more scientifically oriented, generally readily found in the literature, and easy to transfer.
   (b) **Know-how Information:** more technically oriented, not so readily found in the literature, and harder to transfer.
   (c) **Show-how Information:** more operationally oriented (training, maintenance, etc.), rarely found in the literature, and hardest to transfer because it involves not only information transfer per se, but technical and general education as well.
   (d) **Know-who Information:** More personally oriented, reference to sources of information (people, organizations, directories, tables, "invisible colleges", etc.) for asking of specific questions; hard to find, easy to transfer.

3. As to information needed at different stages of development a distinction is made between need for:
   (a) **Information for Preindustrial or Beginning Industrialization Stage,** oriented more toward:
      (i) basic decisions on choices and strategies;
      (ii) diffusion of general knowledge on factors affecting each other in development efforts (e.g. sanitation and health); and

(iii) creation and diffusion of technical and experimental knowledge (know-how and show-how) needed to sustain developmental efforts.

(b) **Information for Relatively Advanced Industrialization Stage,** oriented more toward:

(i) complex decisions, cooperation with other developing countries, negotiation, etc.;

(ii) extension of the technical know-how;

(iii) increase of employment and productivity; and

(iv) diversification, creation of new (and particularly small) industries and new products and markets.

Saracevic and Wood have concluded that as a result of these developments a significant shift is occurring in developing counties as to the perception on needed information. The emphasis is not any more on the traditional international communication in science and science-related technology (know-why information), but on the acquisition and diffusion of knowledge directly appropriate to planning and advancing of national and local, economic and social development. A major characteristic of this shift is emphasis on information that falls into the categories of policy and management information, of operational and industrial information, of experimental and proprietary knowledge, of know-how, show-how, and know-who information—on appropriate information, for short. Consolidated information is an expression of this shift in thinking. It is an attempt to derive and make available appropriate information to various categories of users.

It is important to study the information needs of potential as well as actual users of an information centre in the institution, and any consolidation unit set up within the centre. Knowing their information requirements would involve collection of information on the various research projects they are working on. Such a study, when undertaken, would facilitate the type of collection to be developed and the information services to be designed to meet the information needs of the users.

In order to understand the various fields of interests of individual users to be served by the information centre, it is imperative to know about the following:

3.1 Institutional Characteristics

Some background information need to be acquired about the institution to understand about the information needs of the users. It could be developed in the following possible ways:

(a) Knowing the goals or objectives of the institution, its mission, and projects undertaken. It could be found in some in-house documents.

(b) Knowing the fields of activity and scope of each discernible activity of the persons involved. It can also be found in some internal documents.

(c) Knowing the information they use or need in order to perform their functions efficiently. This can be gathered through such methods like interviews, questionnaire, etc.

(d) Knowing as to where do they get the information they use. This can be gathered through interviews, questionnaires, etc.

(e) Knowing the channels and media through which they get the information. Such information may be gathered through observation of information flow patterns, study of organizational charts, interviewing and questionnaire survey of potential users.

(f) Knowing about the adequate, timely, reliable availability of information, and the cost of its accessibility. This can be gathered through survey, interview, and questionnaire.

(g) Knowing about the interest or awareness of the value of information and of an Information Consolidation Unit among the users.

(h) Knowing if the existing information centre can be modified for a better information consolidation service.

3.2 User Groups

The potential user groups of information and Information Consolidation Unit have to be identified more specifically. This may be done through various methods such as interviews, questionnaires, study of organizational and functional charts, etc. It is important because through such studies the information requirements of a group, and even of individuals within the group, may be determined. In this way it would be easy to determine about the services to be provided to meet their information needs; and the factors which may influence the types of services may include personal attitude, educational background, training, experience, area of specialization, etc.

4. Methods of Survey

There are various methods to study the users, and the selection of method depends largely upon such factors as objectives of study, variables of study, etc. However, the following aspects are involved in selection of methods:

(a) Selection of a sample of user population.

(b) Determination of procedures for collection of data from the sample.

(c) Determination of procedures for analysis of collected data to derive results.

It is advisable to have a detailed look at these aspects before going for the collection of data.

4.1 Sampling

Sampling is important because in most cases it is not possible to reach all the members of a population; it is only a portion that can be practically reached by the researcher. However, that portion should be representative of the entire population so that the results have a high probability of being valid for the whole population.

There are many sampling methods available, but the most common are the following:

4.1.1 Convenience Sampling

It is like picking up the first 25, 50, etc. users that come along as subject of study.

4.1.2 Random Sampling

It is picking up the users for study from a population at random.

4.1.3 Stratified Sampling

This method subdivides the population into subgroups and then picking up the users for study at random from each subgroup.

4.1.4 Representative Sampling

It determines before hand individuals, pairs of individuals, or small groups with some characteristics in common as subject of study.

Each of these methods has its advantages and disadvantages, but none is universally the best. Some are good in certain types of studies than others.

4.2 Size of Sampling

The size of sample to be studied is quite important because a smaller sample is more prone to produce biased results than a larger one. Therefore, a balance needs to be achieved to obtain as far as possible error free results, and also sample should be manageable enough within the limits of time and resources.

5. Collection of Data

The following methods are most often used for collection of data.

5.1 Surveying

In this method the users are questioned and answers obtained directly from them about their bahaviour, attributes, values, conditions, and/or preferences. This is by far the most often used method in user studies.

There may be formal and informal methods of surveying but both depend upon a questionnaire which is most commonly used instrument for data collection. According to

Goode and Hatt, "Questionnaire refers to a device for securing answers to questions by using a form which the respondent fills himself." Therefore, the construction of questionnaire has to be done very carefully to take care in the design of questions. The questions framed should

(a) reflect the variables to be studied,

(b) avoid different interpretation and misinterpretations, i.e., questions should be clear, and

(c) avoid biases which should prejudice answers.

This will help to obtain valid and reliable information to test the hypotheses.

The questionnaire are usually distributed either by mail to a selected sample, or sometimes delivered personally also. When delivered personally, the questions are asked orally in a face-to-face encounter between a user and a surveyor.

5.2 Interview

It is used as a research method by contemporary researchers because the search for new information is not limited to the use of questionnaire only. As the purpose of survey is to acquire current information about the experiences and opinions of users, the interview serves as a useful tool. The interview technique is a verbal method of obtaining data hence it is unique in itself. It is considered as more original because of verbal communication with the interviewer/s.

Young defines interview as "a systematic method by which a person enters more or less imaginatively into the life of a comparative stranger." Goode and Hatt further amplify it by saying that "Perhaps no research technique is as close to the researchers problem as the interviewer."

As compared to questionnaire, personal interview usually not only yields a high percentage of returns but also obtains more correct information.

There are usually two types of interview:

(a) **Structured interview:** It involves the use of a set of predetermined questions set out on a printed schedule and records the answers in the appropriate way.

(b) **Unstructured interview:** This type of interview is characterized by a far too greater flexibility of approach to questioning the users (respondents) as it is informal. The arrangement and wording of questions is determined on the spot by the interviewer.

For greater success, interviews must be conducted in an informal and relaxed manner.

5.3 Observation

It helps to make direct observations on the communication behaviour of users in given situations, practices, time periods, etc. It is a well established method of data collection. W.I.B. Beveridge states that "observation involves noticing something and giving it significance by relating it to something else noticed or already known." Observation therefore is a primary tool of scientific enquiry.

Observation may include aspects such as:

(a) Where users go for various kinds of information;

(b) What sources, structure, channels they use;

(c) What is the frequency of use;

(d) What do they do with answers;

(e) Whom do they inform, etc.

The observation need to be recorded in a consistent manner over time; may be at random times.

As a method, observation is much more difficult than surveys; however, it has the advantage of recording what people do, rather than what they say they do. There may be great difference between the two.

5.4 Records Analysis

In this method the written records or other artifacts of previous communications (such as papers, correspondence, statistics) are obtained and observations about the users are derived from them. These records could be a goldmine for analyzing the communication behaviour, information source used, connections (e.g. by analyzing citations), and so on.

This method is considered as a particular form of observation, and as such it has the same advantage over surveys because it is based on facts rather than opinions. However, it is limited to the availability of records as well as willingness to give access to the surveyor.

5.5 Experimentation

It introduces an element of communication to a carefully defined group of users and observes the results. It can be done only as a formal study. It involves setting up controlled conditions for observation of a carefully determined information source or communication practice. It has a control group and comparison of results between the control and experimental group. Thus, it seems to be the most difficult of all methods.

6. Analysis of Data

Once the data has been collected with the help of the above-mentioned methods, it needs to be analyzed. To make the data useful for various purposes, it has to be summarized, compared, synthesized, tested for significance, etc. Analysis of data is considered to be a difficult task in user studies. There could be an informal analysis of data in which only an impression or feeling of what the data indicate and in which direction they point, is obtained. But for the formal analysis the most often used methods are:

1. **Statistical Analysis:** In this method various standard statistical techniques are used to summarize, compare and test for significance data which is expressed numerically.

2. **Semantic Analysis:** Here, the application of semantic techniques is made to summarize and compare the data which is expressed verbally.

3. **Psycho-Sociological Analysis:** In this method, application of psychological, sociological, or anthropological techniques is made to classify or describe data which is expressed conceptually, logically, or representatively.

4. **Economic Analysis:** Here, the application of macro or micro economic techniques is obtained to derive conclusions in

economic terms on data expressed in either or all of the above ways.

Although knowledge about these methods is required by the surveyors, yet these days standard statistical packages are widely available for use. These packages can accomplish calculations of sums, means, medians, variances, standard deviations, percentages, chi-squares, regressions, or cross-tabulations of data.

References

Atherton, Pauline. *Handbook for Information Systems and Services*. Paris: UNESCO, 1977.

Devarajan, G. *Library Information User and Use Studies*. New Delhi: Beacon Books, 1995.

Goode, W.J. and Hatt, P.K. *Methods in Social Research*. New York: McGraw Hill, 1952.

Kunz, Werner, et al. *Methods of Analysis and Evaluation of Information Needs*. Munchen: Verlag Documentation, 1977.

Ranganathan, S.R. *Reference Service*. 2nd ed. Bombay: Asia Publishing House, 1961.

Saracevic, T. and Wood, J.B. *Consolidation of Information: A Handbook of Evaluation, Restructuring and Repackaging of Scientific and Technical Information*. Paris: UNESCO, 1981.

Scherban, Quoted in Atherton, *op. cit.*

Seetharama, S. *Information Consolidation and Repackaging: Framework, Methodology, Planning*. New Delhi: Ess Ess Publications, 1997.

Steiner, G.A. *Top Management Planning*. Canada: Collier Macmillan, 1969.

Valls, J. "Some Aspects of Information Consolidation in and for Developing Countries." Paper presented at the Third Meeting of UNISIST Working Group on Information Analysis and Consolidation. Kuala Lumpur, 12-16 September 1983.

Voigt, M. Quoted in Atherton, *op. cit.*

Weiner, N. Quoted in Seetharama, *op. cit.*

Whittaker, Kenneth. *The Basics of Library Based User Services*. London: Library Association, 1993.

Arrangement and Presentation of Ideas

4

It is generally observed that the same set of ideas on or about a subject can be arranged in several different ways and sequences. The choice of a particular sequence should be based on the helpfulness to the users which may be the end users of a product. Therefore, it depends upon the purpose and specialization of the users. For example, the information collected for the managers and planners has to be arranged in a manner that facilitates decision-making at their end. Therefore, the choice between different methods of grouping the ideas depends largely upon the orientation required to meet the needs of subject specialists.

1. Guiding Principles

Some guiding principles for the arrangement of ideas are mentioned below:

Perusing the relevant documents on the subject and collection of ideas for consideration of the information consolidation product may be helpful for arranging the ideas embodied in a document.

Therefore, the following may be kept in mind because a document may:

(i) Embody new ideas;

(ii) Give a new interpretation of a known idea;

(iii) Give new data based on fresh observation and analysis;

(iv) Give interpretation of the results of analysis of data to bring out their significance and value;

(v) Present a subject from new perspective, oriented to or establishing correlation between subjects;

(vi) Give historical treatment of a subject;

(vii) Evaluate an existing theory, conjecture or practice;

(viii) Promote or propound a substitute for an existing theory or practice which has been found inadequate; and

(ix) Present known ideas in a new setting.

1.1 Guiding Principles for Ideas

While arranging ideas, one and the same document may contain a combination of two or more of the above categories of ideas. Therefore, the guiding principles for arrangement of ideas will include the following:

(a) Principle of Unity of Ideas;

(b) Principle of Sequence of Ideas; it may further have the following:

1. Principle of Filiatory Sequence;
2. Principle of Decreasing Extension;
3. Principle of Increasing Extension;
4. Principle of Time Sequence;
5. Principle of Development Sequence;
6. Principle of Spatial Contiguity;
7. Principle of Quantitative Measure;
8. Principle of Increasing Complexity;
9. Principle of Concreteness versus Abstractness;
10. Principle of Traditional Sequence;
11. Principle of Literary Warrant; and
12. Principle of Alphabetical Sequence.

1.2 Guiding Principles for Terminology

While translating the ideas into standard terminology, the following guiding principles need to be taken care of.

1. Respecting Unity of Ideas;
2. Principle of Simple Sentence;
3. Principle of Transparency;
4. Avoidance of Tautology;
5. One-to-one correspondence between "Term" and "Idea";
   5.1. Avoidance of Homonym;
   5.2. Avoidance of Synonym;
6. Principle of Technical Terminology;
7. Consistency in Punctuation;
8. Principle of Use of New Term; and
9. Principle of Elegance.

1.3 Guiding Principle for Numbering System

The need to mark the structural elements of the text with a well designed numbering system is being increasingly realized as helpful both to the author and the users. Such a numbering system would indicate the hierarchical system of levels of unity of ideas and the flow of ideas down the structural elements.

1.4 Problems in the Choice of Guiding Principles

There may arise a need for choosing one among the Principles of Helpful Sequence in one or more of the following situations:

1. Two or more principles giving sequences apparently equally helpful in the context;
2. The Law of Symmetry suggesting the possibility of the existence of the sequence of the entries;
3. The need to decide the sequence of application of the principles, when two or more principles are applicable, in arranging the chapters and sections;
4. The possibility of using a principle which will give a comparatively more helpful sequence in the text;
5. The more appropriate attribute of the entity concerned for the application of a Principle of Helpful Sequence.

It is obvious that different user groups of Information Consolidation Product (ICP) may require different sequences of one and the same set of ideas. Therefore, in the choice of guiding principles for arranging the ideas in an ICP, the specific field of specialization, requirement, and orientation of the users should be kept in view.

2. Collection of Information

Before information is consolidated, it is necessary to collect it. It implies the selection and acquisition of information sources keeping in mind the ideas/concepts of the subject concerned. It involves the following steps:

2.1 Identification of Information Sources

- Selection Tools: Bibliographies, selected lists, reviews, reference books, directories, indexing and abstracting sources and services, Selective Dissemination of Information, New titles information, Invisible colleges, etc.

2.2 Selection of Information for Consolidation from among the primary and secondary sources of information

- Identification of selectors.
- Formulation or identification of criteria for selection and evaluation of information and information sources vis-à-vis intrinsic value and user appropriateness.
- Preparation of informative abstracts of relevant documentary and non-documentary sources.

In other words, information and data which are to be readied for presentation could be collected from the following sources:

1. Informal Sources: In our daily life collection of data/ information begins with the generation of ideas through a casual discussion, or reaction of colleagues, friends, neighbourers, or other people. Ideas are also generated in public places, from common official channels, during meetings, personal files, overhearing others, records, etc. Again, for information related to historical aspects, elderly people, village headpersons, senior professionals in organizations, etc. are the best sources of information. These channels help the researchers and others to obtain support from ideas generated from informal sources.

2. Formal Sources: Formal sources of information can either be (i) printed, and/or (ii) electronic.

(i) The printed sources include, as mentioned above,

(a) The primary sources, like journals, conference proceedings, dissertations, standards, patents, etc.;

(b) The secondary sources, like encyclopedias, dictionaries, directories, yearbooks, indexing and abstracting periodicals, atlases, gazetteers, and so on; and

(c) The tertiary sources of information, like bibliographies of bibliographies, etc.

(ii) The electronic sources of information, on the other hand, include the online access databases, CD-ROM databases, Internet access, websites, microfilms, microfiches, and so on.

3. Ephemeral Sources: This category of information sources include newspaper reports, television news, press releases, prospectuses, publicity brochures, booklets, pamphlets, advertisements in print and electronic media, audio and video tapes, etc. They are not of lasting value.

4. Non-Scientific Sources: The non-scientific sources of information are popular magazines, trade journals, public speeches, newsletters, etc.

Information can be collected from any or all of these sources, and, to make it worthy, it should not be collected on piecemeal basis.

3. Organization of Information

Once the required information from relevant sources of information is collected, it needs proper organization for easy use. Organization of information is the basis from which facts and figures are to be arranged in order to produce a quality technical presentation. Information and data should be organized methodically and carefully so that all the good ideas are included in the presentation.

The following points may be considered as important in this respect:

(a) Prepare a single entry for every reference that has been consulted as this facilitates easy reference and quick access for further use.

(b) The definitions, quotations and extracts from the various references should be carefully recorded giving due credit to the author and/or sources. It ensures accuracy and avoids plagiarism.

(c) The source of quotation should be recorded according to some established standard for due credit to the author/s.

(d) The collected information should be arranged in certain predetermined order, or alternatively may be classified according to some classification scheme, or may be arranged under some subject headings arranged alphabetically. This ensures easy retrieval of information.

(e) In order to present information in a desired format it may be reorganized. It will be highly desired if a brief abstract is prepared, and an outline may help to fit the contents of the final document within its framework.

(f) Decisions regarding the introduction, sections, headings, subheadings, paragraphs, facts, analysis, recommendations, conclusions, etc. are to be taken while organizing information.

4. Presentation of Information

The information collected, properly organized, and classified is now ready for presentation for use by users. The need for information by users may be for various purposes such as stimulation (i.e. for creativity and motivation), knowledge (i.e. for competence), current awareness (i.e. for keeping up-to-date with current developments), productivity (i.e. for doing efficiently and effectively), theorizing (i.e. for predicting future phenomena), comparison, problems solving, education , decision making, communication, etc.

It may be mentioned that communication is quite essential in our way of life, but it depends upon the methods of communicating the information or the results of one's work. It has to be decided whether this communication would be written or verbal.

4.1 Parameters of Presentation

In this context it would be better to consider the following parameters:

(a) Level of understanding
(b) Purpose of use of information
(c) Format of receiving information
(d) Subject of the document may decide the presentation technique
(e) Type of information consolidation product
(f) Type of users, etc.

Seetharama has identified the following parameters:

(a) User
(b) Purpose/Impact
(c) Subject
(d) Product
(e) Media of communication—written or verbal

4.1.1 Outline of Ideas

It is, however, advisable to prepare a framework or an outline of ideas to be presented. Such an outline may include the following:

(i) Concise and informative
(ii) Logically organized
(iii) Free from non-relevant terms and concepts
(iv) Readable
(v) Accurate depiction of data and information

4.2 Modes of Presentation

There are broadly two modes of presentation of information, i.e., written and verbal. In case of Information Consolidation Product, written mode of presentation is adopted. The following are the two types of written presentation of information that are more popular.

1. Textual, and
2. Non-textual

While preparing the preliminary draft the use of both the textual and non-textual modes of presentation may be made. After checking and rechecking for accuracies and other technical aspects, the final draft is prepared.

4.2.1 Textual Presentation

This is the simple way of presentation of data and information. However, the presentation of Information Consolidation Product may differ according to the requirements of the user groups. It may be either for the scientists, or engineers, or managers, or policy makers, or planners, or extension workers, etc. Therefore, the presentation of information and layout of the text may change depending upon the target users. For example, in a review meant for the top management the conclusions of the investigation may have to be highlighted and placed at a conspicuous position in the beginning of the report. The detailed data and amplifying information may be placed thereafter so that the top management may apply information/knowledge quickly in decision making. On the other hand, in describing experimental work for scientists, the content elements should be in the sequence: Purpose, Method, Results, Conclusions, Collateral and other information.

The author of the ICP should, therefore, facilitate communication and make use of language accordingly so as to have desired effect on the users. The author should, as far as possible, be conscious of the message to be conveyed when using textual presentation. In other words, the author should use language oriented to the intellectual level, comprehension, and requirements of the users of the ICP.

Therefore, while presenting information in this mode the following aspects may be considered:

(a) The writing style should be strictly impersonal.

(b) The text should have an introductory paragraph indicating clearly the subject of the document.

(c) Only one single topic should be discussed in each paragraph.

(d) While presenting conclusion drawn from several facts, it is advisable to summarize all facts first and then state conclusions.

(e) Simple and direct language should be used.

(f) Long and complicated sentences are not desired, and should best be avoided.

(g) The text should be written in active voice and in present tense.

(h) Care must be taken of the grammar, punctuation, and spellings.

4.2.2 Non-Textual Presentation

It could be presented either in tables, or/and illustrations/ figures. Tables contain only data arranged in systematic order within columns and rows; whereas illustrations include graphic materials like charts, graphs, diagrams, plates, etc. These are discussed as follows:

1. Tables: For research conducted in any discipline of knowledge, data collection and data analysis are the essential characteristics. But to understand fully the complex patterns, relationships, and to facilitate communication, it is imperative that both the tools of analysis and communication are used.

Subjects belonging to sciences, humanities and social sciences are amenable to tabular presentation. However, when information in a Table is further processed for arriving at some conclusions, tabular data must be accompanied by an annotation. Therefore, information given in Tables should supplement, not duplicate the one provided in the text.

While presenting data/information in tabular form, the following points may be considered:

(a) Tables should be self-sufficient, capable of conveying a message independently, and each Table should have a self-explanatory title.

(b) All tables should have numbers consecutively in Arabic numerals, and should be referred to in the text by numbers, and not by terms like 'above', 'below', etc.

(c) Table columns should have brief headings, and for units of measurement standard abbreviations should be used, and these should be placed below the column headings.

(d) Statistical formulae, etc. as cannot be readily typeset inside the tables should be avoided.

(e) Tables should not be made complicated by including too many items and too many details. Rather than making a cumbersome single table, the material may be divided in two or more tables.

(f) The practice of printing big tables, which can neither be shortened nor split up, on a large-sized sheet and folding the same should be avoided because the folds are likely to crack after sometime.

2. Illustrations: Illustrations, referred to as figures also, appear quite frequently in technical books, periodicals, reports, etc. They also appear in the form of drawings, diagrams, charts, graphs, photographs, etc. Illustrations are, however, difficult to make and need professional expertise in this regard.

W.G. Cambell and S.V. Ballou defined an illustration or figure as any kind of graphic representation which cannot be typed or printed.

Illustration should present only one fact or series of related facts and should be easy to interpret. The use of illustration within the presentation is based on the need of and amount of information to be communicated.

(i) **Orthographic Drawings:** These are useful for engineering description and record. These drawings can easily convey all dimensions, specifications, etc. as a guide to the needy user group of mechanists and other production workers.

(ii) **Schematic Diagrams:** These are helpful in illustrating function, path of flow, essential lines of interconnection among components of a system and terminal points. These diagrams can also be used to depict electrical and electronic wiring, architectural piping, transportation routes, system flow, industrial production sequence and flow, etc.

(iii) **Charts and Graphs:** According to Seetharama, these are considered useful as aids in preparing some presentations because these:

- Indicate significant trends of an activity over a period of time. Line graphs are considered useful for presenting a series of data having successive values.
- Provide better management visibility of finances, production, quality control, marketing, and the like.
- Plan and schedule programmes and products through the use of PERT, CPM, and other graphical systems.
- Bar graphs are used for comparing quantities by showing bars of various lengths, horizontal and vertical.
- Show the relationship between the whole and part where each segment of the pie represents one part.
- Represent organizational charts, stages of an activity or stages of manufacture in a factory or process. Flow charts are considered useful here.
- Air programmes of technical training, new employee orientation, and the like.

Nowadays line drawing are computer produced. The advantage of computer application is that it reduces the efforts of reproduction and the drawings can be scanned and included in the text at appropriate places. The size of the drawings can also be adjusted according to the space available within the text of the presentation. The line drawings should be numbered serially in Arabic numerals along with suitable captions. These should be arranged in the text according to the serial order, in which these are cited, for example, Figure 1, Figure 2, and so on.

(iv) **Photographs:** The modern technologies have now made it possible to include photographs within the text. These are considered useful in a presentation. Their usefulness could be as follows:

- To enhance the value of the presentation.
- To authenticate and certify the information contained in the text.
- To make the presentation effective.
- To save time and effort.
- To work as filling material in a technical presentation.

As far as possible, photographs should be produced on glossy paper. Each photograph should also carry an appropriate caption.

The author of ICP has to take a decision regarding the use of textual presentation or non-textual presentation. It largely depends upon the situation to situation and the audience group to whom it is addressed.

5. Technical Writing

For efficient and effective utilization of an ICP, it is necessary that it is written in a simple, precise and clear language. It is important that the ideas arranged in a presentation are properly communicated to the users. But it is generally experienced that most of the scientific and technical literature is written in a language that is followed only by the experts and not others. The language and style adopted instead seem to act as an effective barrier in the communication of ideas and use of scientific literature. It is therefore suggested that this problem can be overcome if the authors of ICPs make use of the techniques of technical writing.

Saracevic and Wood have defined technical writing as a presentation of the essential features of scientific and technical subjects and issues in ways that will make them more understandable and easier to absorb by a wider audience.

They have mentioned that the aim of technical writing is effective application and wider understanding of scientific and technical advances. In other words, technical writing should promote communication by conveying the message concerned. However, N. Roling has cautioned, for this purpose, that enough attention should be paid in formulating messages. He has identified the following functions that messages might have:

1. Drawing attention, or creating awareness;
2. Providing information for assessment;
3. Persuasion by linking promoted behaviour to the needs of target receivers;
4. Instruction, explaining how to do it;
5. Evaluation, helping to assess effect;
6. Improving decision-making capacity, increasing self-confidence, teaching how to handle information and other system-building messages.

In the above-mentioned messages, some are for decision makers, while some others are meant for extension workers. Therefore, messages appropriate to the end users must be communicated through various information consolidated products.

Saracevic and Wood argue that for technical writing to be effective, it has to adhere to principles of good writing in general. They have also offered some specific principles as given below:

1. **Accuracy:** Make it sure that the material selected for technical writing is accurate in accordance to the most up-to-date information available.

2. **Logical Progression:** It is suggested to follow a logical progression of thought.
3. **Simplicity:** Keep it simple.
4. **Word Choice:** Select words for communication, and avoid jargons.
5. **Style and Grammar:** Choose style and grammar for direct and effective communication, not art.
6. **Analogy:** Make more liberal use of examples familiar to audience.
7. **Sensitivity:** Adhere to the sensitivities of the audience, hence follow some restrictions.

References

Arora, Renu. "Technical Writing for Library and Information Professionals". In: *Challenges and Changes in Librarianship: Papers in Honour of Professor Sewa Singh*. 2 volumes. Edited by A. Amudhavalli and Jasmer Singh. New Delhi: B R Publishing Corporation, 2010, pp. 953-74.

Atherton, Pauline. *Handbook for Information Systems and Services*. Paris: UNESCO, 1977.

Campbell, W.G. and Ballou, S.V. *Form and Style*. Boston: Houghton Miffin Co; 1978.

Devarajan, G. *Technical Communication for Information Managers*. New Delhi: Ess Ess Publications, 2012.

Gerson, Sharon J. and Gerson, Steven M. *Technical Writing: Process and Product*. 6th Ed. New York: Prentice-Hall, 2007.

Neelameghan, A. *Presentation of Ideas in Technical Writing*. New Delhi: Vikas Publishing House, 1975.

Roling, N. "Agricultural Knowledge: Its Development Transformation". (Paper presented at the Third Meeting of UNISIST Working Group on Information Analysis and Consolidation. Kuala Lumpur 12-16 September 1983).

Saracevic, T. "Processes and Problems in Information Consolidation". *Information Processing & Management*, Vol. 22 (1) 1986: 45-60.

Saracevic, T. and Wood, J.B. *Consolidation of Information: A Handbook on Evaluation, Restructuring and Repackaging of Scientific and Technical Information*. (Pilot Edition). Paris: UNESCO, 1981.

Seetharama, S. *Information Consolidation and Repackaging: Framework, Methodology, Planning*. New Delhi: Ess Ess Publications, 1997.

Methodology for Information Repackaging

5

For designing an information repackage, the work has to be based on some information consolidated product already prepared. Further, in designing the repackage, it is very essential to have specific information about the target audience to collect, process, and apply required information and design a repackage of information product accordingly.

1. Methodology

The methodology to be followed for this purpose includes the following steps:

1.1 Preparation of the First Brief

The first brief contains selected information on the chosen subject, and is prepared by the information professionals. The repackaged product should give adequate description on the various aspects of the new product, and should communicate to the target audience about its need.

1.2 Analysis of the Brief

The first brief is to be analyzed with reference to the target audience, the information content, the budget of the message carrier as well as the life cycle of the carrier.

1.3 Design Criteria for Message Carrier

The message carrier should be so designed that it attracts the target users. It may, therefore, also include visual information, illustrations, simple textual information with less quantum, etc. It should be in such a way that it looks familiar to the users.

1.4 Selection of the Message Carrier

The message carrier could be designed with several shapes and sizes depending on the requirements of the users. For example, it could be in the form of leaflet, booklet, and so on.

1.5 Designing the Message Carrier

The message carrier in whatever format should be well designed to give it a look of something to be proud of, and feel like owning it. It should be sophisticated enough for its use. Users generally perceive of an ICP, say a booklet, in the form of such things as its shape, size, and cover. In all, it must give a different look to hold the attention of the users.

1.6 Production of the Message Carrier

At this stage all the attention is focused on the execution of the work to produce the message carrier, say a booklet. The actual production is a professional job. Once the design of the product is completed, the rest of the job is done by some commercial firm.

1.7 Feedback

It is important to design a feedback system to judge the success of the repackaged information carrier.

From the above discussion, it may be observed that:

(a) Information repackaging essentially involves selecting, analyzing, and assessing information with a view of communicating a message in a convenient and effective form to a target audience defined for the purpose.

(b) It is very essential to have a thorough knowledge of the target audience, the message, and the message carrier.

(c) The information professional in the project team is required to provide the brief and help the communication professional in analyzing the brief and include visuals, if required.

(d) Well designed information repackages need not be expensive even if visual representations are to be included.

2. Role of Specialist Groups

While preparing the draft text, the author plays the dominant role in the arrangement and integration of ideas into an organized text. The role played by different specialist groups at this stage is given in the following Table 5.1:

Table 5.1: Role of Specialist Groups

S/N	Specialist Group	Type of Role
1.	Subject Specialists/Peer Groups	Provide consultancy in regard to the accuracy, veracity and relevance of information vis-à-vis product. In addition they provide help in draft text preparation as well as in clarifying author's doubts in regard to concepts belonging to fields which he is not familiar. Provide expertise for preparation of supplemental aids.
2.	Library and Information Science Personnel	Provide bibliographic and other standards for presentation of information. Bring to notice of the author, the various principles developed for the arrangement of ideas—principle of unity of ideas, principles for sequence of ideas, guiding principles for verbal plane, etc. in their field. They also help in the preparation of supplemental aids.
3.	Communication and Information Presentation Specialists	Provide consultancy in the choice of appropriate methods and procedures to be adopted in the presentation of information for securing the maximum impact on the product user groups. They would also help in the development of supplemental aids—audio-visual components, slides, cassettes, etc. Editorial help will also be provided.
4.	Intermediaries/Product User Groups	Examine the draft text of the IA&C product from their own view point, i.e. as to how useful the product would be in performing their own functions.

3. Testing of Draft Text

In order to evaluate the quality and effectiveness of an ICP, the draft text is:

(a) reviewed and assessed by subject experts; and

(b) field-tested by the end users or intermediaries.

3.1 Review and Assessment

The ICP may be reviewed and assessed by the subject experts against the following points:

- Technical accuracy
- Completeness
- Perspective
- Analysis
- Synthesis
- Reasonableness of organization
- Appropriateness, utility
- Value-added

3.2 Field Testing

It facilitates improvements in orientation, acceptance and effectiveness of the product in the operating environment.

4. Finalization

The ICP is now finalized, on the basis of the feedback received, after making suggested changes. These changes may be as follows:

- addition or omission of information;
- changes in the sequence of ideas presented;
- changes in the form of presentation (i.e. use of diagrams, graphs, tables, etc.)

After the finalization of the text, it should be properly and carefully edited so as to eliminate any errors, maintain consistency and uniformity in style and presentation.

The manuscript is thereafter handed over to the production department for preparing the printed copies of the ICP for marketing and distributing among the users.

5. Potential Benefits from the use of Consolidated Information Products

In the present era of 'Information Boom' where there is very rapid growth of information/literature, it is imperative to set up Information Consolidation Unit (ICU) in various information centres in order to organize, coordinate, and institutionalize the compression and evaluation activities. This is more important especially in new, peripheral, and interdisciplinary subject fields.

Library and information professionals are well aware of the value of the ICU, but the users and potential users may not be conversant with its benefits. Therefore, Saracevic and Wood have given a list of benefits which could be derived by making use of information consolidated products prepared by ICU. These are mentioned as under:

5.1. The various users of consolidated information and information products get better informed about alternatives and consequences. It helps to reduce uncertainty and improve process of decision making.

5.2. By making use of consolidated information, the users can increase level, breadth and depth of their knowledge. It brings sophistication in establishing relations between unrelated facts.

5.3. Consolidated information facilitates to have more appropriate responses by individuals, groups or organizations to demands and changes in the environment.

5.4. Consolidated information can take organizations to higher levels of outputs in work and other activities. It increases capacity and effectiveness of organizations.

5.5. Information consolidated activities lead to economic use of limited resources, and ensuring increased capacity and/or effectiveness.

5.6. Use of consolidated information will give a wider spread and acceptance of results. Detection of necessary adjustments in the aims of organizations.

5.7. Information consolidation products help the users involved in business enterprises, commerce and related areas.

5.8. Consolidated information helps the industry to incorporate technical innovations, add new products, and make the products more appropriate.

5.9. Information consolidated products help to keep up with research front by making appropriate decisions on policy and allocations.

5.10. Use of consolidated information help to keep up with advances in education, research, methods, etc. It also provides for educational planning and assessments.

5.11. Consolidated information provides for self-fulfillment and advancement in their given area of work. It also provides for wider opportunities in employment and adjustments to changing environment.

It is therefore obvious that for reaping the benefits of an ICP the effective transfer of information to the end users is desired so that they could make its proper use to enhance the quality of work.

6. Criteria for Determining Appropriate Repackaging Activities

We have discussed about the methodology for the preparation of information consolidation and repackaging. However, the discussion on the methodology will not be complete if we do not indicate the criteria for determining appropriate repackaging activities. Broadly speaking, the criteria include the following:

(a) User groups, their characteristics and information needs vis-à-vis their major functions.

(b) Functions/Activities of the information centres.

(c) Availability of resources—both physical and human.

The details about the user groups, their functions and appropriate repackaging activities are given in the following Table 5.2.

Table 5.2: User Groups and Repackaging Activities

S/N	User Groups	Functions	Repackaging Activities
1.	Director, Heads, Middle-level Administrators, etc.	Policy making, Planning, Administration	Subject interest bulletins, literature reviews, State-of-the-art reports, Newsletters, Progress reports.

2.	Curriculum developers, Editors, Subject Specialists, Writers, Trainers, Mass Media Personnel, etc.	Curriculum and Instructional materials development	Actual teaching/learning curriculum materials, guides, Textbooks, Syllabuses, Multimedia packages, Newsletters, Subject interest bulletins, Research abstracts, Data sheets, Handbooks, etc.
3.	Trainers, Trainees, Supervisors, Resource persons, etc.	Personnel and Teacher training	Actual training materials, Audio-visual aids, Multimedia training kit, Training Syllabuses and designs, Newsletters, Research reports, etc. on teaching training programmes, Subject interest bulletins, Data sheets, Handbooks, Yearbooks, etc.
4.	Faculty members, College teachers, Supervisors, etc.	Teaching	Subject interest bulletins, Guidebooks or Handbooks on teaching methodologies, Sample lessons and learning materials, Teachers and pupils guides, Textbooks, Syllabuses, Audio-visual kits, Multimedia packages for classroom teaching, Data sheets, Newsletters, etc.
5.	Researchers, Faculty members, Academicians, Students, Project Evaluators, etc.	Research and Evaluation	Manuals on research and evaluation methodologies, Yearbooks, Data sheets, Handbooks, Research reports, Abstracts, Research journals or bulletins, Newsletters, etc.

7. Manpower for Information Analysis and Consolidation

The work of information collection, organization, etc. and on the basis of this information the creation of an Information Consolidation Product requires the information professional to possess certain knowledge and skills for the purpose. Seetharama has indicated the type of skills required for developing an ICP.

Table 5.3: Information Consolidation Products: Knowledge and Skill Requirements

S/N	Stage	Skill/Ability	Knowledge of
	Idea Plane		
1.	Initiation step –Recognition of need for the Product	To interact with users and recognize their needs	Methods of user studies. Methods for the formulation of a reference frame-work. Methods of data/ information analysis. Highways and byways of subjects concerned.
	–Initiation of work	To identify and organize a team or group of specialists	Methods to be adopted for the preparation of directories of institutions and experts/specialists in the fields of interest of institution or user.
2.	Scope Determination -Subject scope	To specify precisely the subject to be covered by the IA+C product	Methods for study of subjects. Methods for subject analysis/ information analysis.
	-Audience	To specify the product user group and target group	Attributes of user groups and methods for their study.
	-Product type	To identify the products to be prepared for the user groups	Different types of IA+C products and their preparation to suit specific user needs.
		To identify the subject specialists/peer groups for consultation	Methods to be adopted for the preparation of directories of experts specialists in the fields of the institution.
3.	Collection of Information	To identify sources for selection of information	Methodology for identification and preparation of a list of core documents, specially core periodicals.
		To identify and select relevant information	Subject concerned Abstracting/extracting/ condensation techniques.
		To technically process the information	Classification/indexing/ analytical techniques.
4.	Appraisal of Information Collected	For evaluation of information	Evaluatory techniques to be adopted. Criteria for assessing the quality or intrinsic merit of

			information—validity, reliability, accuracy, credibility, significance, etc. Specific subject concerned and methods for study.
5.	Scope Re-definition	To redefine the scope in consultation with end users, intermediaries, etc.	As in Stage 2.
6.	Draft Text Preparation	To consolidate and integrate the information into an organized text	Consolidation and integration techniques, including packaging and repackaging. Guiding principles for arrangement of ideas. Guiding principles for the Verbal plane. Guiding principles for the Notational plane. Technical writing. Different methods of presentation-tables, graphs, illustrations, etc. Bibliographical and other standards for presentation of information. Supplemental aids and their preparation.
7.	Draft Text Testing	For evaluation of IA+C product	As in Stage 4 In addition, knowledge of feedback techniques to be employed with the intermediaries/ beneficiaries.
8.	Finalization	To incorporate the changes suggested To provide aids to the users of the IA+C products	Standards to be followed in the preparation of aids.

8. Sum-up

It may be concluded up from the above discussion about the methodology for the preparation of Information Analysis and Consolidation Products that these can be prepared in both textual and non-textual format depending upon the requirements of the users. Many such ICPs in the form of

Handbooks, Manuals, etc. have been prepared in India also. According to Seetharama, the following points may be taken into consideration while preparing ICPs.

1. If the objective is to prepare both information packages (e.g. Handbooks) and repackages (e.g. Manuals) then general-to-specific model should be adopted as it is more helpful.
2. The function/activity—oriented model or reference frameworks would facilitate preparation of ICPs for decision-makers and extension workers as well as for other non-discipline oriented users.
3. In most cases, ICPs are the result of the joint and integrated effort of various specialist groups.
4. In developing countries, Information Consolidation Centres with necessary and sufficient infrastructural facilities can prepare and produce ICPs like handbooks and manuals oriented to the needs of their users in a most economical way.
5. Information consolidation activities are complex and need efficient and effective information manpower.
6. Information consolidation products are to be oriented to the different user groups if they have to be familiar and useful.
7. While preparing the ICPs, the user of the product concerned should be kept in mind.
8. ICPs cannot be either totally textual or totally non-textual. Since a symbiotic relationship exists between the two types, presentation of information in a product should be based on an optimal mix of text, tables, illustrations, etc.
9. One should develop the capacity and capability to supplement words by using tables and figures as well as to use words to focus attention on the significant features of the tables and figures.
10. The principles and practices of visual communication have a direct application in information repackaging

as they contribute substantially to the effectiveness of information repackages.

References

Atherton, Pauline. *Handbook for Information Systems and Services*. Paris: UNESCO, 1977.

Documentation Research and Training Centre. *Annual Seminar*, 18. Bangalore: DRTC, 1981.

Saracevic, T. "Processes and Problems in Information Consolidation". *Information Processing & Management*, Vol. 22 (1) 1986: 45-60.

Saracevic, T. and Wood, J.B. *Consolidation of Information: A Handbook on Evaluation, Restructuring and Repackaging of Scientific and Technical Information*. Paris: UNESCO, 1981.

Seetharama, S. *Information Consolidation and Repackaging: Framework, Methodology, Planning*. New Delhi: Ess Ess Publications, 1997.

Weisman, Herman M. *Information Systems, Services and Centres*. New York: Becker and Hayes, 1972.

Abstracting: Types and Guidelines for Preparing Abstracts

6

The importance of fast communication of results of research has long been realized by the researchers, scientists, academics, etc. to keep themselves abreast of the latest developments. This is particularly true in scientific and technological research where current information has to be made available efficiently and effectively.

In this context, the growing importance of indexing and abstracting services, as an important link in the communication chain is well known to the scientific community the world over.

An abstract helps in the selection of relevant documents in current and retrospective literature search on a given subject.

ABSTRACT

1. Definition

An abstract is an abbreviated, accurate representation of the significant contents of a document. It is usually accompanied by an adequate bibliographical description to enable the original document to be traced.

An abstract is the summary usually by a professional (abstractor) other than the author, of the essential contents of a work, usually an article in a periodical, together with the specification of the host document and the locus within it.

In other words, an abstract is a concise representation of the contents of a document, in a style similar to that of the original document.

The American National Standards Institute has defined an abstract as an abbreviated, accurate representation of the contents of a document, preferably by its author for publication with it. In other words, it is a concise condensation of the significant content of a document presenting its objective, scope and major findings.

2. Related Terms

Some other terms are also used to denote the condensation or abbreviation of a document's contents. These are defined below:

Annotation: It is a note added to a title or other bibliographic element of a document by way of comment or explanation.

Extract: It is composed of one or more portions of a document that have been selected verbatim to represent the whole.

Summary: It is a restatement within a document (usually at the end) of the document's salient findings and conclusions; it is intended to complete the orientation of a reader who has studied the preceding text.

3. Functions

An abstract performs the following functions:

(i) Facilitates selection of documents to be read completely by providing in a summarized form the ideas/concepts, dealt within the original document;

(ii) Functions as a mechanism by which the specialist can keep abreast of developments in the field of specialization as well as other related/interested fields;

(iii) Facilitates in the preparation of subject bibliographies, and state-of-the-art reports, etc.

Good abstract must be acceptable and adequate from the view point of the users. H. Borko and S. Chatman have specified that an adequate abstract of a document must contain information on the following:

(a) Purpose of writing the documents;
(b) Methodology/procedure followed with a statement of experimental techniques used;
(c) Results, i.e. a statement of the findings; and
(d) Conclusions, i.e. a statement dealing with the interpretations or significance of the results.

4. Purposes

An abstract serves the following purposes:

(a) To provide the reader with all the essential information, like the nature and purpose of the work, new approach, novel findings, results and conclusions, in order to enable the reader to decide whether or not to consult the original document;
(b) To transmit important information to readers in a concise but precise manner;
(c) To promote current awareness;
(d) To save the time of the readers;
(e) To help overcome language barrier;
(f) To facilitate literature searches.

5. Characteristics

An abstract must be brief and accurate, and presented in format designed to facilitate the skimming of a large number of abstracts in a search for relevant documents. It is well defined type of literature with definite attributes and a unique style. It must possess the following characteristics:

5.1 Brevity

Abstract should be shorter than the document from which it is prepared. Therefore brevity enhances its value. An abstract is read very carefully by the readers so as to draw conclusions about the original document from every word or sentence. Brevity saves the time of readers, saves space, and brings in pinpointed information to the readers.

5.2 Accuracy

An abstract should be as accurate as possible because it leads the user to the original document. Accuracy is required with regard to writing the citation of the original document and the body of the abstract. The mention of data details and the conclusions provided in the abstract should be absolutely correct. In case of any inaccuracies, the users are likely to be misled. Element of accuracy will increase its use, respect, and confidence in the abstracting service.

5.3 Clarity

Abstract should be written clearly in a style that is easy to read and understand. For this purpose, it is best to write, as far as possible, in author's own words and in complete sentences. Writing in abstractor's words sometimes changes the meaning of the author. If abstractor is preparing an abstract, it is essential that he paraphrases the author's work and not extract information, as paraphrasing enhances the quality of the abstract.

6. Types

There are different types of abstracts that are prepared by different people for different purposes. Some of the types are discussed below:

6.1 Informative Abstract

It provides a concise summary of the significant contribution made by the document. It represents the principal information (including results, conclusions and recommendations) underlying the topics. A good informative abstract is a miniature of the original work, and it has a beginning, middle, and an end with emphasis on key ideas and results. It usually saves the users the necessity of consulting the original document.

6.2 Indicative Abstract

It generally indicates the main topics discussed in the original document, but does not contain much of the information; hence it cannot be used in place of the original. This type of abstract is useful for describing lengthy texts and is used for representing multi topic texts to indicate or describe what can be found in the original document. It is thus written for enabling the user to decide whether he should read the original or not.

6.3 Critical Abstract

A critical abstract is an informative abstract where the abstractor not only describes the contents of the document but also evaluates the work by including his opinion and analysis. The value and quality of this kind of abstract depends upon the subject competence of the abstractor.

6.4 Slanted Abstract

A slanted abstract is also referred to as special purpose abstract. It can be informative, indicative, critical, or short abstract that is oriented to the interests of a particular readership. It is usually written to comply with objectives of the abstracting service that publishes it. Therefore, in such a case, for a single document, two different abstracts are prepared for two different services belonging to different subjects as they are slanted to their needs.

6.5 Modular Abstract

In this case, the abstractor reads the complete document before abstracting, and prepares a set of abstracts rather than just one abstract, with a just little additional time. Each of such modular abstract can be used with a different purpose or abstracting service. The modules to be prepared may include informative abstract, critical abstract, annotation, or indicative abstract.

6.6 Short Abstract

A short abstract, comprising of two or three sentences, is valuable in commercial information services and in news sheet type of information dissemination for public services.

6.7 Telegraphic Abstract

Such an abstract is usually prepared with the purpose of brevity by selecting significant words from the document. It sounds somewhat like a telegram so why it is given the name telegraphic abstract. In this style of writing, grammatical or function words and even verbs are eliminated, leaving a mere listing of topics. Its main objective is to make it helpful for computerized searching while still maintaining the human intelligibility. The telegraphic abstract is thus a self-indexing abstract prepared manually so that computer could search abstract as it were an index.

6.8 Mini Abstract

As it is a highly condensed, brief abstract, it is referred to as mini abstract. Mini abstract is a sort of indicative abstract which provides a line or two about the document. It works well for documents that have good explanatory titles and are used along with current awareness services for notifying information. Such an abstract is used in commercially significant, highly competitive and dynamic technical fields of knowledge.

6.9 Statistical or Numerical Abstract

This type of abstract is used for certain specialized subjects and scientific papers where emphasis is on presenting data. Statistical abstract usually summarizes the results in a data table which can easily be reproduced with the citation.

6.10 Author Abstract

It is the abstract prepared by the author of the document and forms a part of the original paper. It may be an indicative or informative abstract depending on the nature of publication. Although abstracting requires training and experience, authors generally are the best persons to write abstracts of their papers as they know the subject well. Most of the international abstracting periodicals use author abstracts with modifications, if necessary.

6.11 Subject Expert Abstract

It may be an indicative or informative abstract prepared by a peer in the specific subject of the document either by education or specialization and experience. It is prepared when the authors do not prepare the abstract themselves, or depending on the policy of the abstracting service.

ABSTRACTING

Abstracting is the process of preparing abstracts. Basically abstracting is a writing style which is achieved with good command of language, thorough knowledge of the subject and is improved with practice. Although abstractors attempt to write in a clear, terse, accurate and non-critical manner in a style similar to the original document, yet different abstracts may be written for the identical documents depending upon the readership.

In order to make the process of abstracting as uniform and consistent as possible some abstracting guidelines and published criteria are developed from time to time. They suggest including in the abstract at least the objective of the document, the method used, the results, and the conclusion.

7. Guidelines for Abstracting

Guidelines for abstracting are considered essential for the abstractors so that they could prepare abstracts with a degree of uniformity. Therefore, agencies, or organizations, or publishers involved in the preparation of abstracts for various kinds of documents, usually prepare a set of guidelines or instructions for the abstractors.

Jennifer Rowley has developed some guidelines for preparing abstracts relating to the following points:

1. Style and length of abstract.
2. Instructions on the way in which abstracts are to be prepared, for example, typing, paper, layout, spacing.
3. Language of the abstract and treatment of foreign language materials.
4. Abstracting procedures.
5. Special approaches to abstracting for various types of materials.
6. Form of bibliographic reference or citation.
7. List of standard abbreviations including those of journal titles and standard forms of quoting chemical nomenclatures and mathematical expressions.

8. Introduction to the abstracting service and an explanation of the nature and use of abstracts.
9. Indexing instructions, especially if the abstractor and indexer is the same person. These instructions may include index terms and classified headings or notations.
10. Sample forms for competing and submitting abstracts, including instructions for conformity with computer processing.
11. Examples of abstracts in various fields of interest to the abstracting organization.
12. The criteria to be used in selecting items for abstracting, for example, subject, form, source.
13. Transliteration and the practices to be adopted when citing items in languages with non-Roman alphabets.
14. Proofreading procedures and marks.

However, S. Seetharama has suggested certain guidelines for preparation of abstracts in more details. These are given below:

7.1 Purpose

The purpose of guidelines is that the statements pertaining to the process of abstracting must be comprehensible and adaptable, particularly if they are to prove useful.

7.2 Objectives

The aims of providing these guidelines are:

(a) to identify the essential items of information to be included in the abstract, and

(b) to suggest a method for the preparation of an abstract.

7.3 Need

The need for providing these guidelines is:

(a) to maintain uniformity and consistency in abstracting;

(b) to improve the quality of abstracts; and

(c) to conserve the time and energy of abstractors.

8. Steps

These guidelines are in the form of steps which would be useful in the preparation of abstracts of both macro and micro documents.

The different steps involved in abstracting are discussed below:

8.1 Selection of Material for Abstracting—Phase I

The abstractor must bear in mind some points in selecting material for abstracting. Some of the Do's and Don's in selecting material are given below:

Table 6.1: Do's and Don's in Selecting Material

S/N	Do's	Don's
1.	Selecting relevant material/documents keeping in view the interests of the clients/users.	Do not select material which is irrelevant to the clients/users.
2.	Selecting significant material.	Avoid trivial material.
3.	Select original and novel contributions.	Avoid repetitious (non-original) material.
4.	Select documents	Avoid selection of documents
	i. Published from reputable sources	i. Published in sources of unknown or questionable reputation.
	ii. Less accessible, i.e. those which involve considerable difficulty in procuring. Eg.: Documents of foreign origin especially societal and association publications.	ii. Readily accessible, widely circulated domestic articles.
	iii. Published in sources as abstracting service has committed itself to cover by policy.	iii. Published in sources usually excluded by policy.
5.	Select final reports and other reports which are adequately supported by data or evidence.	Do not select preliminary progress reports inadequately supported by data or evidence.

8.2 Abstracting Proper—Phase II

It involves the steps as given below:

(i) **Identification of the Focus:** The first step in the preparation of an abstract is the identification of the focus (central theme) of the subject embodied in the document. Identify the primary advances, if any, and secondary advances in knowledge as well as specialized information in case of particular subject fields. For the purpose of identification of focus, the document has to be read completely, especially where the author states the purpose/objective, method, results, and conclusions.

(ii) **Content Analysis:** In this step, the abstractor analyses the information content recognizing the relationships between the concepts discussed in the document and notes them down without omitting any of them. The language used should be precise and concise, that is, free from any verbosity.

(iii) **Assessment:** The third step in this phase involves the work of assessment of information collected (in the previous step) for its usefulness. This is a very important step as it would determine what should be included in the abstract. Information selected should be beneficial to the users. One of the purposes of assessment is to achieve a desired degree of brevity in the abstract.

(iv) **Synthesis:** The abstract is prepared by synthesizing in a cogent and coherent manner the information collected as a result of the previous steps.

(v) **Checking:** On the completion of writing the abstract, the abstractor should check again for errors, omission of information, etc. This is a very important step in that if the reader is surprised by what he finds in the original, he will have no faith in the abstracting service.

(vi) **Editing:** Many of the abstracts will not be suitable for publication in spite of the use of various rules and criteria of abstracting. It is said that less than 1% of the abstracts prepared are suitable for publication without editing. The reasons may be either deviation from the criteria or omissions and in some cases errors. Editing also helps in maintaining consistency and uniformity in style and presentation.

9. General Rules/Criteria

Seetharama has provided the general rules/criteria pertaining to structure, language, style, etc. for preparing the abstracts. These should be familiar to the abstractor before he starts abstracting a document.

(a) The abstractor should be aware of the users/audience to whom the abstract is meant and their informational requirements. This would facilitate orientation or slanting of the abstract to the user/s.

(b) An informative abstract should generally be provided for original articles from periodicals, conference proceedings, patents, technical reports, etc. Indicative abstracts may be provided for review documents, standards, theses and dissertations, etc.

(c) The abstract should indicate the type of document—book, article, review, state-of-the-art report—and also whether the work is theoretical or experimental.

(d) Use simple, precise, and concise statements but with some variety in length and structure to avoid monotony. Use complete sentences. Telegraphic language can also be used, but without sacrificing precision or clarity.

(e) Use phrases for clauses, words for phrases when possible.

(f) Use standardized terminology, abbreviations, acronyms, initialisms, symbols, etc. in the abstract.

(g) Use numerals for numbers when possible.

(h) Avoid generalization in statements.

(i) Do not interpret the author's statements/remarks.

(j) Avoid both overusing of and awkward omission of articles.

(k) Avoid labeling each part of the abstract.

(l) Paragraphing in abstracts is not desirable.

(m) Each abstract should consist of three sections or parts. These are:

  (i) **Reference Section:** Directs the readers/users to the original document.

  (ii) **Body Section:** Contains data from the original document and/or indicates or describes the content of the original.

  (iii) **Signature Section:** Identifies the abstractor.

(n) The Reference Section is essential for precise guidance from the abstract to the original document. Hence, it must be complete and accurate. The following items of information must be provided in the Reference Section as they are considered significant.

  (i) Name of Author/s.

  (ii) Title of the document.

Note: Ambiguous titles should either be made clear or ambiguity should be corrected in the Body of the Abstract.

Foreign language titles to be reproduced as such with an English translation.

  (iii) Locus of the document. Information of locus includes:

    – Name of the Title of Periodical in abbreviated form.

    – Volume number, Issue number.

    – Year and month of publication.

    – Inclusive pagination.

Note: In case of a macro document, the name of publisher, place and year of publication, number of pages, and price are to be given.

Other items of information that may be provided, if relevant, are, Accession number of the document (if a macro document), name of series, Author affiliation,

Agency sponsoring the work, Contact/Grant number, Report number, alternative sources for procuring the document, etc. The sequence in presentation of the above-mentioned items of information can be decided in consultation with the clients/users. When there is controversy regarding the sequence, other methods, such as use of italics, use of boldface, etc. can be adopted.

(o) The Body Section should contain in a condensed form the thought content of the original document. It consists of two parts—DESCRIPTORS and ABSTRACT PROPER. The sequence between the two parts depends on the purpose for which descriptors are given. If the descriptors are considered as a brief guide to the contents, they should precede the abstract proper. On the other hand, if they are considered as distracting, they can be placed after the abstract proper. In either case, the descriptors have to be selected using a standardized vocabulary as in a thesaurus, classification scheme, subject heading list, etc. The assignment of descriptors is a very important step as they facilitate storage and retrieval of information.

(p) The Signature Section should contain the name of the abstractor. The signature may be either full name or initials.

10. Special Rules/Criteria

The special rules/criteria pertain mainly to the Body Section of an abstract. In the preparation of the abstract, two processes are mainly involved—selection of content for inclusion in the abstract, and arrangement of content. It may be observed that there is some amount of overlap in some of the steps which needs to be taken care of.

11. Conclusion

The guidelines given above are based on the generalization and experience gained during the preparation of abstracts for technical articles by the various abstractors. Most of the guidelines available seem to be tailored to the immediate needs of the various abstracting agencies.

In view of lack of some standardized and universal set of guidelines for preparing abstracts. However, the International Organisation for Standardization (ISO) has prepared a standard *ISO 214 Documentation—abstracts for publication and documentation 1976*. It provides a useful outline on the basic principles of writing informative abstracts, and is used by some abstracting agencies as guidelines for preparing abstracts for their services.

In India, the Documentation Committee of the Indian Standards Institute (now Bureau of Indian Standards) took up the responsibility of preparing standards for abstracting way back in 1953. These were published as IS 795-1956 under the title *Canons for Making Abstracts*. However, in view of the current requirements these need to be revised to make them more comprehensive and explicit in stipulating standards for different types of abstracts.

References

Ashworth, W. "Abstracting", In *Handbook of Special Librarianship and Information Work*. 4th ed. London: ASLIB, 1975.

Borko, H. and Chatman, S. "Criteria for Acceptable Abstracts: A Survey of Abstractors". *American Documentation*, Vol. 14, 1963: 149-160.

Guha, B. *Documentation and Information: Services, Techniques and Systems*. 2nd ed. Calcutta: World Press, 1989.

Kumar, P.S.G. *Information Analysis, Repackaging, Consolidation and Information Retrieval*. New Delhi: B R Publishing Corporation, 2004.

Rowley, Jennifer. *The Electronic Library*. 4th ed. London: Library Association Publishing, 1998.

Seetharama, S. "Guidelines for Abstracting". *Eighth IASLIC Seminar*, held at Madras, 1978: Papers. Calcutta: IASLIC, 1978.

Information Consolidation Products

7

1. Introduction

After ascertaining user requirements through user studies, the results may be used in developing information consolidation products that will satisfy their needs in given subjects. The information consolidation products are determined with one eye on the user and the other on the best available information on the subject. As user needs vary so do the information consolidation products (i.e. tangible outputs of information consolidation units) because they relate to the user studies conducted to know their information requirements.

2. Definition

Information consolidation products are defined as those that are prepared after selecting and acquiring suitable information sources. These are those products that emanate from information consolidation units (ICUs) of various organizations. These products involve the performance of various activities like selecting acquiring, evaluating, analyzing, synthesizing, restructuring, packaging, repackaging, etc.

Seetharama has suggested that while preparing information consolidation products the following points may be kept in mind:

(a) A good information consolidated product is the result of the joint and integrated effort of different groups of specialists.

(b) The users of the information consolidation products may not be the end users (beneficiaries) of the information contained in the product, but may be intermediaries whose role is to cull out from the product the pertinent information and present it to the beneficiaries in a convenient form.

(c) Information consolidation products may be a by-product or spin-off from the work done (or services provided) for producing other information products/services.

It may be mentioned that the information consolidation products include information provided in either electronic or printed form. But they are not the same as the physical products. Physical products are generally prepared or manufactured for some profit purposes, while information products are meant to provide the latest information on a subject scattered in a variety of information sources. Therefore, information products represent a different value proposition.

Information based products have very high development costs, but very low cost per unit manufactured. For example, preparing a database on a CD is quite capital intensive to create, but the cost of reproduction is insignificant. Once a master copy is developed, generation of additional copies of the product costs almost nothing.

3. Types of Information Products

Once the information needs of users have been identified they must be met by preparing and producing suitable information consolidation products (ICPs). There are different types of ICPs prepared to meet with different requirements of users. Some of the types of ICPs, as enumerated by Saracevic, are given below:

(a) **Reviews:** These include critical reviews and state-of-the-art reports.

(b) **Reports:** These include assessment, technical reports, market reports, alerting bulletins, newsletters, and house journals.

(c) **Data:** This category includes data compilations and tables, statistical abstracts, correlations, composites, and critical data.

(d) **Databases:** These include expert databases and subject knowledge bases.

(e) **Technical Writings:** This type has in its fold guides, manuals, instruction sheets, explanations on a complex subject provided in a style and language geared toward a given audience level, popular articles on a scientific or technical topic.

(f) **Handbooks:** These include compilations of essential data, and information on a subject.

(g) **Critical Studies:** This type includes comparison of different practices or policies, or studies, and impact and future studies.

(h) **Requests:** These include briefings, evaluative on-demand studies, and compilations.

It may be pertinent to mention that each of the above stated products has its own requirements, criteria, procedures, etc. For example, criteria and practices for reviews are quite different from those for technical writings.

4. Content Aspects

From the above discussion, it becomes clear that many products will require more or less restructuring of synthesized information. Restructuring requires the determination of a mode and language in which the contents of information consolidated products (ICPs) will be presented to users in a way that is most compatible with their needs.

The following aspects need to be considered in relation to contents in a given ICP.

4.1 Content Extent

It implies the decision on the degree of detail with which a subject is treated in a given information product. It involves such dimensions as subject breadth, i.e. general or broad presentation to specific or narrow; and subject depth ranging from summarized or selective to detailed or comprehensive treatment.

4.2 Technical Sophistication

It defines the technical level and understandability of the contents of an ICP. It includes:

(a) minimal information with a single or very few concepts with no technical jargon;

(b) moderate level with several concepts, and uses some technical terminology;

(c) high level makes use of almost all essential concepts with liberal use of technical terminology; and

(d) very high level includes discussion of complex relations with other topics or subjects, and makes liberal use of concepts and terminology from other subjects.

4.3 Temporal Aspects

It describes the two basic time dimensions involved, i.e., immediate period (short period, such as six months); or current period (within the last year or so); or retrospective time (prior to a year and extending to a defined historical period). It also describes about the frequency of issue of the ICP (i.e. if it is to be one time only, or periodic issues, periodic update, etc).

4.4 Editorial Qualities

It includes a number of qualitative aspects associated with the style of presentation. The prime function of this facet is to make the contents more palatable to users and easier to assimilate. Therefore, it involves such aspects as grammar, syntax; clarity, crispness; balance, contrast; logical organization, coherence; layout; graphic qualities, composition, colour, etc.

4.5 Information Quality and Accuracy

It explains the degree of known validity and reliability, i.e. the quality of information or data presented in the ICP. Therefore, it involves evaluation of information prior to consolidation in a product.

4.6 Degree of Invariability or Value Added

It defines the degree to which information is presented as contained in the original text/s. When evaluation is performed the degree of value added in terms of points of view, comparisons, results of tests, judgments, etc., is considered.

4.7 Degree of Linearity

It implies the degree to which the order of presentation in the product is the same as the order in the original text/s.

5. Design and Development

In the present era of rapid and enormous growth of information in a variety of formats, it has become well nigh impossible for the users to cope with the information overload. Information products and services play a critical role in helping users by making available to them information in desired formats so as to meet with their information needs.

There could be the following distinct concepts of a product:

(a) **Tangible Product:** It is the physical entity that is offered to the customers. For example, books and periodicals, camera, biscuits, soaps, computer, etc. are all tangible products. The tangible products, however, may have some such characteristics as styling (i.e. distinctive look or feel), features, quality, packaging, and branding.

(b) **Expected Product:** It is the tangible product along with the whole cluster of services that accompany it. It includes a set of attributes or features which a customer normally would expect from it. For example, in an institutional library, the document collection is expected to focus on the objectives, mission, vision, and areas of interest of the institution.

(c) **Augmented Product:** It refers to the attributes that meet the desires of the customers beyond their expectations. For example, provision of a photocopy facility for library users and installing an online access catalogue for them are like augmenting the library products.

(d) **Potential Product:** It includes the augmentations and transformations that can happen in the product offerings in the future. Organizations must continuously develop products in view of the technological changes, market dynamics, etc.

However, information professionals are concerned with information products which are designed and developed for the users to meet with their information requirements.

As discussed already, data and information are collected from a variety of information sources depending upon the type of product to be designed and developed.

The development of information products, according to Kotler, involves the following steps to be taken in this process:

1. **Idea Generation and Acquisition of Information:** The new product process starts with the search for ideas. New product ideas can come from interacting with various groups or people and from applying idea and creativity-generating techniques.

   The required information and data from the relevant information sources has to be collected mostly by the information professionals. It required due consideration for the quality, scope, coverage, depth, accuracy, timeliness, relevance, credibility, cost, control, and so on. The sources of information must be of standard and high quality so as not to compromise with the ultimate quality of the information product.

2. **Idea Screening:** Once the ideas have been generated, these are properly screened at this stage so as to know their probable acceptability. Therefore, the unattractive and unrealistic ideas are eliminated.

3. **Concept Development and Refinement:** It involves conversion of idea into a sound concept acceptable to customers, and refinement takes place by value addition to the selected one. Refinement can be

physical (e. g. conversion of data from one medium to another) or logical (e.g. labeling, indexing, integrating, etc.). Refining may also include cleaning or standardizing data. Again, meta-analysis on repositories may also be performed, for example, using statistical packages to add value. Therefore, refining adds value to the repository not only by creating usable information but also by enabling information to be stored flexibly. This would help in generating various information products efficiently.

4. **Marketing Strategy:** Once the concept has been developed and refined to the desired extent, the manager of the new product will develop a tentative strategy for introducing the product in the market. The organization also takes decisions regarding the viability and cost of the product and/or service.

5. **Product Development:** On the basis of the target customer requirements a prototype of product and/or service is developed with the help of quality function, customer and other technical attributes.

   This information may then be stored on magnetic or optical media so that it could be easily retrieved at a later stage with the help of software meant for the purpose.

6. **Market Testing:** The new product and/or service is introduced in the market with a view to know the market reaction in view of customer expectations.

7. **Distribution:** It represents the form in which the information product is delivered to the end user. The distribution may be in the form of print, fax, CD ROM, etc. Distribution encompasses not only the medium of delivery but its timing and frequency as well because both are equally important.

8. **Presentation:** One of the basic characteristics of information products is that the value of information, and therefore, the value added in producing information products are pervasively influenced by the

context of their use. Ensuring ease of use and sufficient functionality is part and parcel of the information product itself. It has been experienced by many information products companies that delivering a file of information on CD ROM, for example, is not nearly as valuable as giving a user an easy to use interface to navigate through and interact with the information as per requirements.

These steps are only generic ones. They are likely to vary depending on the type of information product/service to be developed. For example, steps for a consolidated product such as a trend report or state-of-the-art report will be different from those of a compilation of a product such as a directory.

The above-mentioned steps are shown in the following Figure 7.1.

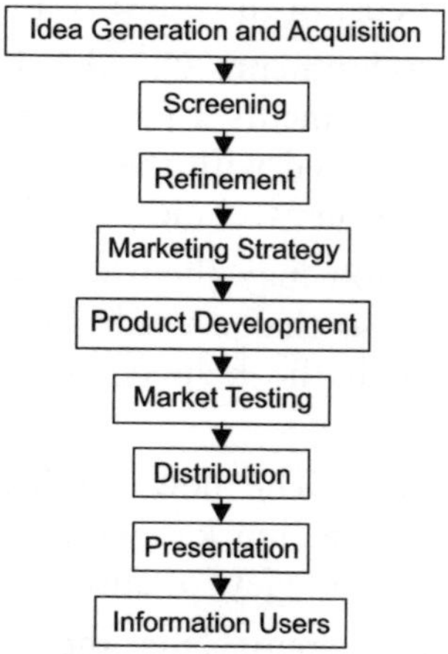

Fig. 7.1: Information Product Development

9. Packaging/Repackaging of ICP

This is another important aspect in preparing an Information Consolidation Product, discussed briefly here. It involves presenting the restructured information in a form that will enhance the potential of its use. Saracevic considers two aspects in packaging or repackaging, i.e. the media and formats in which an Information Consolidation Product could be presented to users. He opines that the major issue is:

which media and formats should be used in order to enhance the potential use of an information consolidation product and contribute to the comprehension, assimilation and recall of its contents by the users?

9.1 Package Media

A number of package (physical) media are available in which information is displayed or presented to users. The new media are being constantly developed and therefore choices are wide.

Basically, the media can be grouped into print, microform, audiovisual, and electronic media, as well as interpersonal contacts. Each of these, according to Saracevic, has different requirements, criteria, procedures, and technology for creation, and most importantly, different economics.

9.2 Package Formats

It refers to the shape and make-up of the physical presentation of an ICP. Like the different package media, numerous package formats are also available to be used for arrangement and presentation of consolidation information in any of the media. Specific formats are clearly dependent on the specific medium used for a given product.

However, the choice of a particular format should reflect not only the physical medium but also (i) the content and nature of information to be presented, (ii) the users and their environment, and (iii) the means or channels of distribution of and user access to information.

Thus, decision on packaging media and formats should be taken quite carefully.

10. Diffusion Stages

As has been discussed above, the design and development of Information Consolidation Product (ICP), depends largely upon the information needs of the users. According to the view of Saracevic and Wood the ICPs would be related to the information needs in different stages of the diffusion process. They have mentioned five stages or phases into which diffusion can be partitioned, viz. Awareness, Interest/Knowledge, Attitude Formation, Trial/Decision, and Adoption/Confirmation.

The role and purpose of the products at each stage differ considerably; they may be described as follows:

10.1 Awareness

It exposes the users to the general existence of an innovation or idea. At this stage there is a need to create general consciousness through examples, and guide the users to the next and more substantive sources of information.

10.2 Interest/Knowledge

It awakens further interest. Providing basic details of the innovation or idea (e.g. technology), and creating a cognizance of the knowledge, know-how and steps needed for trial and adoption. There is a need to guide the users to sources that will stimulate further learning.

10.3 Attitude Formation

At this stage relating the innovation or idea to user needs and functional answers within users' situation is accomplished. The users are informed about the general utility of the idea or innovations. Users must be given enough details to compare this idea or innovation to others in their own environment. Here arises the need to guiding the users to information sources that will generate appraisal, trial, and decisions and persuading them to do so.

10.4 Trial/Decision

It provides applied details to measure performance or applicability of the innovation against requirements. The users are assisted to obtaining evaluations, i.e. performance data on the specific utility of the innovation to their particular situation. Users are directed to sources that facilitate trial and decisions. Also, they are guided to sources for additional technical or operational details.

10.5 Adoption/Confirmation

This is achieved by providing continuing information to satisfy requirements of the modified environment of users, i.e. as changed after adoption.

It is clear from the above, as it passes through the five stages, i.e. from awareness to confirmation, the information requirements become more specific. In turn, as ICPs become more specific, the effort and cost involved would increase correspondingly.

Saracevic and Wood have derived, from the model of Herbert Landau et al., the examples of ICPs as they are appropriate for a number of different types of user organizations, *and* as they relate to the five stages in the diffusion of innovations or ideas. These are presented in Table 7.1.

Table 7.1: Products by Diffusion Stages

Type of Organisation	Awareness	Interest/ Knowledge	Attitude Formation	Trial/Decision	Adoption/ Confirmation
Small Industry	Advertising announcing reports on a particular product/ process. Promotional brochure. Newsletter. Data.	Reports. Group seminar. Visits. Briefings. Exhibit. Synthesis of requirements.	Seminar on effects. Comparative statistics. Report on alternatives and experiences. Videotapes, films.	Demonstration. Technical manuals. Graphic displays. Data books. Handbooks.	Operational manuals. Market reports. Handbook.
Medium Industry	Brochure describing product/process of interest to technical personnel. Market description for managers.	State-of-the-art review describing technology. Resource and economic analysis or summary.	Comparative experiences. Critiques. Market analysis.	Demonstration. Technical assistance manuals. Films, tapes pictures. Data books.	Operational manuals. Updates. Trends. Synthesis of advances. Handbook.
Large Industry	Alerting bulletins. Newsletter. Technical news synthesis. Market trends.	Individual seminar on new developments. State-of-the-art. Economic analysis. Synthesis of research results. Critical data. Reply to inquiry.	Competitive analysis. Impact analysis. Historical review of similar efforts. Evaluations.	Demonstration. Testing reports. Technical manuals. Training manuals. Tables of recommended data.	Operational manuals for different levels of personnel. SDI synthesis of advances. Continuous critical and recommended data.
Local Government	Promotional brochure. Press release. Newsletter announcing reports on topics of local interest.	Workshop for officials. Reviews. Statistical synthesis. Visits. Analysis/ synthesis of practices.	Impact analysis. Comparative data. Demonstration. Briefing.	Manuals for implementation. Comparative experiences. Referral. Documentation.	Recurring supply of synthesized information. Establishment of question-answering procedures.

Type of Organisation	Awareness	Interest/ Knowledge	Attitude Formation	Trial/Decision	Adoption/ Confirmation
State Government	Announce-ments. Form letters. Presentation at meetings. Newsletter. Alerting bulletin.	Slide/tape descriptions. Briefing. Reviews. Statistical composites. Synthesis of research findings. Data tables.	Historical data. Comparison with other countries. Demonstra-tion. Impact analysis. Evaluations. Integrative future studies.	Synthesis of practices and trials. Manuals. Referral. Documenta-tion.	Recurring supply of synthesized information. Evaluation reports.
Mass Urban Population	Newspaper articles. Announcem-ents on radio, TV. Popular talks on a given practice or product. Question-answer sheets.	Radio, TV, shows. Popularization articles. Simplified pamphlets, brochures. Comic strips. Photo novels. Demonstr-ations. Exhibits. Popularized journals. Field manuals.	Comparing with other practices. Demonstra-tions. Synthesis of fitting with daily life. Possible impacts, positive and negative.	Show-how information. Simplified manuals. Do-it-yourself booklet.	Synthesized information fortifying and refining the practice.
Mass Village Population	Radio announce-ments. Visits with opinion leaders. Posters, graphic displays. School talks. Training sessions for extension workers.	Extension service demonstration. Simplified pamphlets. Field manuals in various media (e.g. picture books, comic strips). Films. Talks.	Talks by opinion leaders. Comparison with other villages. Synthesis of fitting with village life. Impacts. Summary.	Show-how information. Field manuals. Experiential information. Markets for cash crops. Do-it-yourself booklets.	Reinforcement information. Comparisons. Refinements. Increased know-how information. Market enlargement information.
News Media	Press release. Invitation to demonstra-tion. Press clippings from foreign or technical press. Alerting bulletin.	Demonstration. Popularized articles. Synthesis of findings and results elsewhere. Timetables. Plans.	Graphic material. Comparative data. Briefings and news conferences.	Short guides. Interviews with experts, leaders.	Reports on applications. Articles on refinements. Result analysis. Impact analysis.
Academic/ Research Institutions	Journal notices. Short announce-ments in professional newsletters. Acquisition notices.	State-of-the-art and critical reviews. Synthesized research results. Referral. Journal articles. Text excerpts. Critical data. Data survey.	Synthesis of critical reports and evaluations. Comparative results. Resource requirements information	Research data. Training/ education materials. Know-why information. Experiences at other institutions.	Continuing synthesized SDI reports. Selective current contents. Refinements in advances. Critical data.

11. Dissemination of ICP

With the preparation of an ICP or provision of a service, the work of an Information Consolidation Unit is not completed. This Unit has to ensure that an effective dissemination of the ICP is made so that there is effective communication and fruitful use of the information contained therein. In the absence of dissemination the products will remain on shelves and the services will be unused.

Some of the more important channels for dissemination of ICPs are as follows:

1. **Interpersonal Delivery:** It includes the products provided personally to users either at their request or in anticipation of need.
2. **Group Personal Delivery:** It means products delivered to a whole group of users such as in a meeting, demonstration, etc.
3. **Strategic Placement:** It has products placed in locations frequented by users to pick on their own.
4. **In-House Delivery:** It indicates that an institution of distinct dissemination functions in an organization such as circulation, reference, etc.
5. **Local Depositories:** It is cooperative arrangement with local information systems and libraries.
6. **Mass Media:** With the help of mass media such as newspapers and magazines, as well as through broadcasting (radio and television) the information products are delivered and/or announced for the benefit of users.
7. **Mail:** This is yet another channel of disseminating information products directly and through mass mailing.
8. **Telephone:** This channel is used either through hot lines for mass uses or direct interpersonal contact with possibility of feedback.
9. **Computer Networks:** These are used for delivering information product via telecommunication networks.

It includes online searching of databases, e-mail, computer conferencing, etc.

10. **Various Combinations of Channels:** These can be used as per the requirement.

It is therefore incumbent on the library and information systems to actively involve in the dissemination of ICPs. Yet time and again, as Saracevic rightly observed, information systems throughout the world—both developed and developing countries—neglect to pay adequate attention to active and effective dissemination.

12. Product Life Cycle

The product life cycle is based upon the biological life cycle. This is true in case of information products also. Information products, therefore, pass through four stages, viz., introduction, growth, maturity, and decline. These four stages, according to Kotler, are discussed below:

1. **Introductory Stage:** As the new product is introduced in the market there is slow sales growth. It requires promotion and need to inform the potential customers/users. For the physical goods also the profits will be low at this stage.
2. **Growth Stage:** At this stage there will be further increase in the acceptability of the information product. In the case of physical goods also the sales will pick up and the promotional activities will continue. The company may use at this stage several strategies to sustain rapid market growth.
3. **Maturity Stage:** The acceptability of the product continues, it reaches saturation, and enters a new stage of relative maturity. This stage lasts longer than the previous two stages in the life of an information product.
4. **Decline Stage:** In case of physical goods also a regular decline in sales begins at a particular stage. There may be a number of reasons for this decline including technological advances, shifts in the customer/user requirements, etc. This is a stage of introducing an

alternative product to meet with the changing needs of users.

These four stages of product life cycle are illustrated in the following Fig. 7.2.

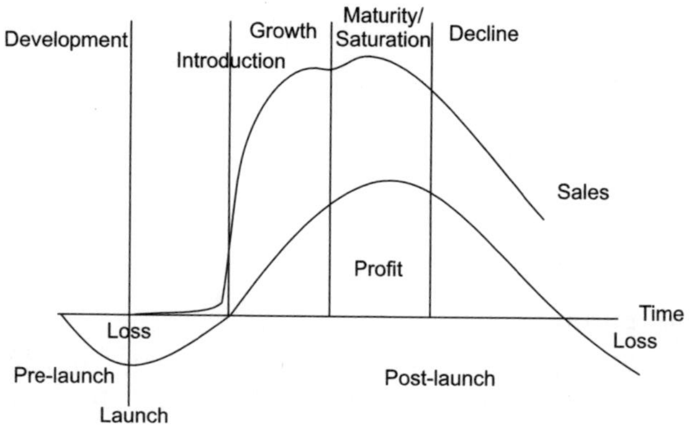

Fig. 7.2: Product Life Cycle

Source: Lancaster and Withey.

References

Atherton, Pauline. *Handbook of Information Systems and Services*. Paris: UNESCO, 1977.

Cronin, Blaise. *The Marketing of Library and Information Services*. London: ASLIB, 1981.

Kotler, Philip and Keller, Kevin L. *Marketing Management*. 12th ed. New Delhi: Prentice-Hall of India, 2007.

Lancaster, Geoff and Withey, Frank. *Marketing Fundamentals*. New York: Elsevier Science and Technology, 1994.

Landau, Herbert B., *et al.* "An Information Transfer Model to Define Information Users and Outputs with Specific Application to Environmental Technology". *Journal of the American Society for Information Science*, Vol. 33(2) March 1982: 82-91.

Saracevic, T. "Processes and Problems in Information Consolidation". *Information Processing & Management*, vol. 22 (1), 1986: 45-60.

Saracevic, T. and Wood, J.B. *Consolidation of Information: A Handbook on Evaluation, Restructuring and Repackaging of Scientific and Technical Information*. Paris: UNESCO, 1981.

Seetharama, S. *Information Consolidation and Repackaging: Framework, Methodology, Planning*. New Delhi: Ess Ess Publications, 1997.

Marketing of Information Products and Services

8

1. Introduction

In recent times marketing has become much more important and significant, particularly in business and industrial organizations than what it was earlier. These days marketing has become a function of management. However, marketing is different from selling. One sells an automobile, but marketing is looking at transport needs and offering a vehicle that satisfies a transport need. One sells books and periodicals, and even information, but markets a solution to a problem. Again, one offers/sells an information consolidation product, but looks at a market and devises a product that the market wants.

2. Concept

Marketing is, therefore, an aggregate of many activities such as promotion, advertising, etc. The central concept in marketing is in being customer and not product oriented. The modern concept of marketing is, therefore, used to identify the customer, satisfy the customer, and keep the customer. With the customer as the focus of its activities, marketing management is one of the major components of business management. It is an integrated process through which companies build strong customer relationships and create value for their customers and for themselves.

However, there is a common misconception that promotional activities constitute marketing. This is not true as rightly pointed out by Saracevic and Wood who define marketing as an aggregate of many activities, promotion or advertising being one of them.

Philip Kotler, a well known authority on marketing has defined it in its new context as "the analysis, planning, implementation, and control of carefully formulated programmes designed to bring about voluntary exchanges of values with target markets for the purpose of achieving organizational objectives. It relies heavily on designing the organization's offering in terms of the target market's needs and desires, and on using effective pricing, communication and distribution to inform, motivate, and service the markets."

The UK Charted Institute of Marketing defines marketing as "the management process that identifies, anticipates and satisfies customer requirements profitably."

A different concept is the value-based marketing which states the role of marketing to contribute to increasing shareholder value. In this context, marketing is defined as "the management process that seeks to maximize returns to shareholders by developing relationships with values customers and creating a competitive advantage."

Blaise Cronin, on the other hand, defines marketing as a management style based on a service philosophy with the following characteristics:

1. User/client—orientation rather than product/ materials or process-orientation.
2. Concern for quality rather than quantity.
3. Emphasis on effectiveness rather than efficiency.
4. Stress on the benefits to be derived by the user/client rather than the intrinsic merits of the service on offer.

M.J. Zachert and R.V. Williams observed that "marketing is planning that focuses on products, place or mode of delivery, adjustment of cost/price to the market, and promotion to specifically targeted segments...."

The interpretation of marketing has been changing with the change in times. According to Theodore Levitt it is now being viewed in a broader context as "a customer-satisfying process, not a goods producing process."

American Marketing Association has defined marketing as "the process of planning and executing the conception, pricing, promotion and distribution of ideas, goods and services to create exchange that satisfy individual or organizational objectives."

Another commonly used definition of marketing is that it "is a social and managerial process by which individuals and groups obtain what they need and want through creating and exchanging products and value with others." In both these definitions marketing is explained as a process related to an exchange and marketing management is the set of activities that facilitate this exchange.

Thus, the essence of marketing involves finding out what the users want, then setting out to meet those needs. With the customer as the focus of its activities, it can be concluded that marketing management is one of the major components of business management to provide "the right product, in the right place, at the right time, at the right price."

From the foregoing definitions it can be inferred that the concept of marketing is quite dynamic, and therefore valuable to all the organizations engaged in the production of goods and services.

In the context of libraries and information centres, marketing means sufficient change in the traditional attitude of librarians towards acquisition, organization, processing and retrieving information. The basis of library and information services should be to help its users to solve their information gathering and processing needs. It must therefore adjust its products, services, organizational policies and procedures to the requirements of the users.

3. Need for Marketing

In recent times, the need for marketing of goods and services has gone beyond the traditional advertising, selling, distribution, etc. Marketing has become an academic subject and makes extensive use of social sciences, psychology, sociology, economics, etc. and has evolved into a profession in itself.

Marketing has spread to nonprofit and public sector organizations such as educational institutions, government programmes, museums, and even libraries and information systems. Kotler argues that marketing need was felt and marketing philosophy has been successfully to nonprofit and public organizations.

It is considered essential for the following reasons:

3.1. The libraries and information centres invest huge funds on collection, processing, and storage of information resources. It is usually observed that these resources are put to very meager use which implies, in a way, wastage of precious funds. In view of the increasing cost of information there is need for the promotion and use of information resources, hence marketing is required.

3.2. The needy users have casual or almost nil need perception because of which they do not use and demand information. In view of the inadequate resource utilization, information marketing is essential to facilitate need perception of the needy and thereby create demand.

3.3. Information is considered as the fifth need of man; hence users must be conscious of their needs for information. Lack of realization of this aspect puts one to disadvantage and deprivation of proper information use. Marketing of information will eliminate these lacunas and make the needy an information rich user.

3.4. The information providers must not confine themselves to their corners; they must rather come out of it and publicize effectively their products and services to ensure their optimum use.

3.5. As marketing creates and increases demand for information services and products, the image of information centres and status of information providers would improve.

4. Benefits of Marketing

Those involved in the marketing activities do so to make the marketing system efficient and effective. To achieve this aim, marketing has to be a well developed programme. The following benefits can be obtained from such a marketing programme:

(i) Improved satisfaction of the users.

(ii) Extension of service to potential users.

(iii) Efficient use of marketing resources.

(iv) Improved resource attraction to the organization.

It is therefore, clear that within the limited resources available at the hands of the organizations they need to extend their user base so as to make the efficient and effective use of resources.

5. Functions of Marketing

It is an established fact that marketing is a management function by which business organizations create customer interest in their products and services. It includes such tasks as analysis, planning, implementation, and control. These are described below:

5.1 Analysis

The foremost task before the marketers is to undertake analysis of the customers, markets, and competitors. Such an analysis helps to investigate specific market and customer needs to target the appropriate customer groups.

Besides, it is important to know the strengths and weaknesses of an organization by using the SWOT analysis. The strength (S) of an organization lies in its ability to satisfy customers; while weakness (W), being the internal factor reduces the ability to meet the market requirements. Opportunity (O) lies in the potential to improve the ability for customer satisfaction; while threat (T) can reduce the ability of the organization.

5.2 Planning

Under the planning stage setting of objectives, making choice of target markets, marketing strategies, etc. are covered. Development of long range marketing plans, new products and/or service development, etc. are included in this stage.

5.3 Implementation

This stage includes such activities as recruiting staff, allocating tasks, making budgets, and organizing activities.

5.4 Control

Progress is evaluated against objectives and targets in this stage. But if need arises corrective measures are also taken to exercise control over the marketing as follows:

(a) Performance standards are set

(b) Progress of marketing goals is measured

(c) Results compared with the standards

(d) Changes are made to stay on track.

6. Marketing Strategy

Broadly speaking, a marketing strategy is a process that can allow an organization to concentrate its limited resources on the greatest opportunities to increase sales and achieve a sustainable competitive advantage.

According to Philip Kotler, a marketing strategy should be centred around the key concept that customer satisfaction is the main goal.

Marketing strategy is a method of focusing an organization's energies and resources on a course of action which can lead to increased sales and dominance of targeted market niche. A marketing strategy combines product development, promotion, distribution, pricing, relationship management, and other elements. It identifies the firm's marketing goals and explains how those will be achieved within a stated time frame.

Marketing strategy of a product is the formulation and implementation of marketing programme to support the perspective of strategic marketing.

It determines the choice of target market segments, positioning, marketing mix, and allocation of resources. It is most effective when it is an integral component of overall firm strategy.

Marketing strategy essentially deals with interplay of three forces known as the strategic three Cs, viz. the customer, the competition, and the corporation (i.e. organization). Most marketing strategies focus on ways in which the corporation can differentiate itself effectively from its competitors by delivering better value to its customers.

6.1 Marketing Strategy in Library and Information Centres

As the marketing strategy focuses on user needs and wants the producers also produce such products as would be meeting their needs. To develop a marketing strategy the organization would use certain analyses and tools some of which are given below:

6.1.1 Market Analysis

The market analysis will help in knowing the potential markets where exchange may take place. For library and information services/products, the market constitutes of end users like the clients, the parent institution, the government, and other funding agencies. Once this market is identified, its size may also be determined which is likely to affect the demand for information services and products.

Segmentation of the total user market is also a part of the market analysis. It will help in knowing the distinct needs and wants of each segment and designing the services/products accordingly.

6.1.2 User Analysis

User analysis will help the library staff to know about library users in detail. It will be established as to which user group or/and individuals shall be making use of information services/products. The users can even be further categorized to make the services/products more useful to them.

6.1.3 Organizational Analysis

Library and information centre, being an organization in itself, knows well its market of users. Accordingly, it designs its services and products to offer to different user segments. This analysis would be based on the available resources, attitude of authorities, expertise of staff, and so on. Such an organizational analysis helps the library to know its strong points which it may ultimately turn into opportunities of marketing its services/products.

6.1.4 Competitive Analysis

Library and information centre has also to know like any other organization about its potential competitors in other similar institutions. Libraries also compete with one another in getting a reasonable quantum of funds, and further providing an array of information services. While undertaking competitive analysis the user segment as well as information services/products of so-called competitor must be evaluated. Such an analysis will help to compare user needs and wants, and thereby designing of products and services.

7. Elements of Marketing

Saracevic and Wood have identified that marketing of information consolidation products and services involve at least the following elements:

7.1 Market Research and Analysis

It implies research into analysis of information needs and wants of a population in terms of segments of potential users, their communication patterns and habits, economic and other constraints, and other characteristics which influence the choice of an information product or service. This would include suggestions on alternate information products and services that would satisfy needs and wants of each segment. It would also include evaluation of usage of product or/and service. Thus market research, as any other research, is a combination of (i) assumptions, (ii) facts, figures and observations, and (iii) human imagination and judgement.

7.2 Product Development and Targeting

Development of a product has to be targeted to the needs and wants of a given segment of users. It also involves consideration of alternatives, determination of the criteria and procedures for selection and evaluation, analysis and synthesis, and overall production and dissemination. (This has already been discussed in an earlier Chapter.)

7.3 Costs and Pricing

As all information processing involves costs, there is no such thing as "free" information therefore, in the present times of market economy, costs and pricing of information products and services need to be worked out. Cost of information refers to all expenditures incurred in production and dissemination of information products and services. There are various pricing schemes that can be applied for this purpose. These include marginal cost, cost recovery, past prices, variable price, market will bear price, costs plus profit, etc.

7.4 Promotion

Promotion of an information product or/and service should be considered in terms of communication to and with specific user segment, and not just an announcement. Promotion can be done in many ways such as by brochures, pamphlets, posters, guides, handbooks, advertising, publicity, exhibitions and demonstrations, and even personal approach, and so on. Promotion should be a continuous activity based on an optimal mix of different approaches.

7.5 User Education

The aims of user education according to Saracevic and Wood, are to:

1. Communicate about a product or service and show how it fits in a broader framework and concern of users.
2. Explain the basics of its structure and function, limitations and problems.
3. Explain in detail its benefits.

4. Explain in detail the way it can be accessed and used.
5. Explain clearly direct and hidden costs, requirements, conditions of use.
6. Elicit comments and evaluations from users for improvement of the product or service.

The main role of user education is to ensure the credibility of the product and trust, in its worth and to increase understanding, know-how and skills in using information.

7.6 Dissemination

Dissemination pertains to the conveying of information to users through given channels, and includes the spreading about, distribution and delivery of information products and/or services. Some of the important channels for dissemination include:

1. Interpersonal delivery
2. Group personal delivery
3. Strategic placement
4. In-house dissemination
5. Local depositories
6. Mass media
7. Broadcasting/Telecasting
8. Mail
9. Telephone
10. Computer networks

7.7 Evaluation of Products and Services

Evaluation of products and services of an information consolidation unit is required on a continuing basis. It should be based on users' criteria placing value on the following:

1. **Quality** of information provided (i.e. precision, accuracy, credibility, recency).
2. **Scope** of information provided (i.e. completeness, comprehensiveness, and coverage of topic).

3. **Appropriateness** to user needs (i.e. fitting the need, level, language, sophistication, information overload, understandability, ease of use, etc.).
4. **Hassle** in getting the information (i.e. time lag, paper work, ease of access, red tape, etc.).
5. **Costs** (i.e. direct price paid, hidden costs, etc.).

8. Marketing Mix

8.1 Concept

The concept of marketing mix can be defined as a process related to an exchange, and marketing management is the set of activities that facilitates this exchange. The marketing activities have been traditionally depicted as marketing mix which has been defined as a set of marketing tools an organization uses to pursue its marketing objectives.

E. Jerome McCarthy classified these tools into four broad groups, which he called the four Ps of marketing: product, price, place, and promotion. Marketing mix is regarded as the key concept to perform marketing functions. But according to Elliot De Aze, marketing mix is the planned package of elements which will support the organization in reaching its target markets and specific objectives.

Neil H. Borden is said to have started the use of the term "marketing mix" in 1964. But Borden derived the term "marketing mix" from James Culliton who described marketing executive as a "mixer of ingredients" who prepares plans to satisfy the target customers. Borden liked the idea of calling market executive as a 'mixer of ingredients'. An executive is "a mixer of ingredients, who sometimes follows a recipe as he goes along, sometimes adapts a recipe to the ingredients immediately available, and sometimes experiments with or invents ingredients no one else has tried." It was indeed nothing but "marketing mix".

Borden identified a set of twelve elements in marketing mix which McCarthy refined further and defined the marketing mix as a combination of all the factors at a marketing manager's command to satisfy the target market. Thus

McCarthy, in his classical model, regrouped Borden's twelve elements into four elements or the 4Ps.

However, there have been divergent opinions among scholars about the number of Ps. Some intend to add one more P to the existing 4Ps, some 3Ps, or some even suggest as many as 15Ps in marketing mix. In this backdrop Chai Lee Goi expresses serious doubts about the role of marketing mix as a management tool in its original form and raises the need for more Ps in the present environment.

It may, however, be mentioned that marketing mix is a powerful concept because it makes things easy to handle, allows the separation of marketing from other activities of the organization, and the delegation of marketing tasks to specialists.

8.2 Elements of Marketing Mix

Meanwhile, we can discuss about the traditional 4Ps, viz. product, price, place, and promotion. These elements need to be blended artfully to get maximum benefit out of it. In the present day client oriented market philosophy, the four Ps can be translated into 4Cs of marketing as shown below:

1. Product ---------------- Customer needs and wants
2. Price ---------------- Cost to user
3. Place ---------------- Convenience
4. Production ---------------- Communication

1. Product

Product is the first and most important element of the marketing mix. Kotler said that "A product is anything that can be offered to a market to satisfy a want or need." These products include physical goods, services, experiences, events, persons, places, properties, organizations, information, and ideas. The product offered to the customer is for attention, acquisition, use or consumption that would lead to the satisfaction of want or need.

(**Note:** It has already been discussed in detail in Chapter 7).

2. Price

In the emerging market economy, information is not coming for free. All information processing involves costs, according to Saracevic and Wood, but all information products or services may not involve a price for a customer. They define cost as referring to "all expenditures incurred in production and dissemination of information products or services." They further stated that "Pricing is the process of determining the price to be charged to the users."

Price, therefore, is the only element in the marketing mix that generates revenue for organizations. But price is all around us. We pay rent for a house, tuition for education, fare for train travel, wages for a worker, honorarium for a lecture, and so on. From the customer view point price is cost or the amount charged for a product or service. Price is the perceived value that is exchanged for a product and/or service, etc. The customer, therefore, pays the price for the benefit of using a product or a service.

In marketing process, price fixation is an important part. The market manager has to take a decision regarding setting for a new product so that it can face the competitive market environment. Product price can be based on two types of factors, viz. internal and external.

2.1 Internal Factors

Many internal factors, as given below, have impact on the price of a product.

(i) Cost of the product is one of them which means total cost that includes fixed and variable costs. Fixed costs generally do not change even if the volume of production undergoes a change. But variable costs change proportionately.

(ii) Marketing objective is another important variable for price fixation. If the objective is survival in the market the company can stay as long as it covers fully the variable costs and a part of fixed costs as well. If the objective is market leadership, a low initial price may be enhanced later.

(iii) Marketing mix strategy is another internal factor that affects price decision. Sometimes price of the product is fixed after giving due consideration to other mix elements like place, promotion, etc.

(iv) It is important as to who is fixing the price. In some organizations the price is fixed by the management, while in some others product managers take pricing decision.

2.2 External Factors

Some external factors also influence the pricing decision for a product.

(i) Nature of market or demand for product has its influence on the price. Therefore, the relationship between demand and supply should be analyzed.

(ii) Competitors' costs and price offers of substitute products should be taken into account while fixing the price.

(iii) Other environmental factors like economy, industrial boom or recession, inflation, and government policies may affect price decision of the products/services.

However, pricing of services sometimes poses difficulty particularly in the context of library and information services. In order to overcome this difficulty, the service providers have based the price on the quantity of inputs rather than service output. But library and information services have been free of cost hitherto; hence there is further difficulty in the absence of any pricing objectives.

In the emerging market economy library and information professionals are also considering pricing their services and products. There is a general feeling that:

(a) customers would not realize and recognize the value of the service and/or product unless it is priced;

(b) the government is impressing upon more and more organizations to become financially self-supporting as far as possible.

2.3 Pricing Policy

Basically there are three types of pricing policy followed for products/services:

(i) **Cost Based Pricing:** It is based on either total or marginal cost of production. Mark up pricing and cost plus pricing are common cost oriented pricing methods. Price is determined in both by adding fixed percentage to the unit cost of production.

(ii) **Demand Based Pricing:** The intensity of demand, not the levels of cost, is taken into account. In case of weak demand lower prices are charged; while higher price may be charged for more demand. This method is called price discrimination.

(iii) **Competition Based Pricing:** The price is not decided on either the demand or cost of production, but on the ongoing price in the market. It is popular pricing method because it represents the collective wisdom of the industry. This type of pricing is possible only in case of homogenous products or services.

2.4 Pricing Objectives

Pricing techniques adopted by any organization depends upon their goals and objectives. Some of them may be as follows:

(i) **Cost Recovery:** An important objective is to recover a portion of the production cost. In case of library and information centres, the fixed costs may be paid by the host institution or the government, and the operational costs or variable costs by the users.

(ii) **Market Penetration:** While introducing new products or services in the market, maximum participation pricing is often used. In order to attract the customers low prices are fixed for the product or service.

(iii) **Market Limitation:** When a product gets established in the market, low prices may lead to excess demand, and the prices need to be raised to balance the demand.

(iv) **Market Equity:** It helps to fix the price according to the economic nature of goods, i.e., public goods or private goods. In case of private goods the benefit goes to the individuals, while the public goods are funded by the public taxes and not through user charges.

Thus, there may exist several goals and objectives, but the library and information centres with diverse users and services/ products may follow multiple pricing objectives.

3. Place

The third element of marketing mix is 'Place', i.e., where the product is made available. It is most commonly referred to as distribution channel or intermediary. It is the mechanism through which the products/services are moved from the manufacturer or the service provider to the user or consumer. So, all activities related to placing the products at the 'right place' are covered under this element. 'Place', which should be according to the convenience of the consumers, comprises of management of distribution channel, physical distribution of products and/or services.

For proper distribution, it is important that regular and smooth flow of products/services is maintained so that supplies are not delayed. The physical distribution of products/services involves design of satisfactory storage, transportation and delivery, inventory control, order processing, etc.

There are two types of elements in 'Place' that need to be examined, viz., distribution channels, and physical distribution.

3.1 Distribution Channels

The marketing managers use distribution channels to display or sell their products/services to the buyers. These include distributors, wholesalers, retailers, and agents which work as intermediaries or middlemen. Involvement of intermediaries is important as they not only reduce the problems of producers in distributing the product over a large area but also give feedback to producers for further improving the product. However, the more the intermediaries, the higher the price of the product the end consumer pays. The most common distribution channel followed is where the product moves along its path from producer to the wholesaler to consumer in supermarkets. The choice of distribution channels also depends upon such factors as the nature of product, nature of market, intermediaries, size of company, marketing environment, etc.

3.2 Physical Distribution

It includes all such activities which help the manufacturers to supply right type of product at the right place at right time. Some components of physical distribution are as follows:

(i) **Order Processing:** Customers' orders for the product are processed with computer application by the manufacturer and therefore carried out quickly saving time and money.

(ii) **Warehousing:** Storage facilities are important for a proper flow of the product because production and consumption cycles generally do not match.

(iii) **Inventory:** It is generally desired by marketers that the manufacturers must have enough stock ready to be supplied in the market. But as it involves heavy cost it may not be possible, hence they have to plan it very carefully.

(iv) **Transportation:** The physical distribution of products over a large geographical area has been made easily possible by various modes of transportation. The choice of right type of mode of transport does affect the cost and condition of products.

By following this distribution channel a manufacturer ensures that his products are easily accessible through stores to target customers.

On the other hand, some marketers do not follow the steps in the distribution channel and instead market directly to buyers through direct mail, telemarketing, door-to-door selling, shopping via television (a growing trend in the present times), or factory outlets.

3.2.1 Place and Library

Place is very much significant for library and information centres as well. If a library is not located at a place convenient to its users its services are not likely to be used most effectively. Library and information centre should be located as far as possible at a central place convenient to all the users. In case it is not so located it must emphasize on the need to design promotional activities for its services and products.

But in the present computerized environment, howsoever inconveniently a library is located its online services and sources are likely to be easily accessed by users.

Similarly, an electronic database can be accessed and used anywhere in the world provided there is network connectivity.

But a public library is location bounded and its branch libraries have to be located across the town/city, and deliver services to users. Regarding the easy and fast access to library services and resources, in addition to 'Place', other important components are opening hours, rules for lending documents, help from staff, etc.

4. Promotion

Promotion is the fourth important marketing mix element. It refers to communication with the customer. It is necessary for the marketing manager to communicate with the intention of informing and using persuasive tactics to change the preferences of customers for the products/services. In the modern marketing environment he needs to build relationship with the customers and promote products with the following objectives:

(i) informing, at the early stage, about products and services available;

(ii) quality of products being offered;

(iii) benefits the products can provide to customers; and

(iv) persuading the potential customers to use the products.

Various methods of promotion form part of promotion mix. It is like a cake mix where the basic ingredients are always the same. But if we vary the amount of one of the ingredients, the final outcome is different.

There are some major tools that make up a promotional mix. Kotler and Keller have listed such tools as sales promotion, advertising, sales force, public relations, and direct marketing. These are discussed as follows:

4.1 Advertising

Advertising includes all forms of paid, non-personal communication and promotion of products, services or ideas by a specified sponsor. Thus, it is a technique paid for communicating about the product. It is impersonal one way mass communication process and gives reason to buy a product. Advertising appears in such media as print (newspapers, magazines, billboards, flyers), in-house newsletters, professional society journals, or broadcast (radio, television), cinema, outdoor advertising (such as posters, bus/rail sides). The advantage of advertising is that it reaches a large number of people. However, it does not provide any feedback.

4.2 Sales Promotion

It offers a short-term incentive designed to encourage immediate purchase of a product or service. Sales promotion often attracts those customers who switch from one brand to another, looking for low price and good value. It includes such activities as displays, exhibitions, demonstrations, discounts, rebates, free gifts, etc. Thus, in market where brands are highly similar, sales promotion can help increase in sales, but it may not be a permanent increase. The use of sales promotion has been popularized since the late 20th century due to such factors as increased sophistication in sales promotion techniques, greater pressure to increase sales, rise in number of brands, decrease in the efficiency of traditional advertising, etc.

4.3 Personal Selling

Personal selling is oral, face to face communication with one or more prospective buyer/s, making visits and presentation in homes or in fairs, for the purpose of selling a product/service. In the present times, personal selling is considered most effective tool for promoting the sale of a product and/or service and also building customer relationship. This method is deemed to be more persuasive, hence sales representatives can convince customers about the utility of the product/service.

4.4 Public Relations

Earlier it was called as publicity, but public relations in the context of marketing goes beyond simple publicity and has importance in many ways. In contrast to advertising and sales promotion public relations involves less commercialized modes of communication. Its primary purpose is to disseminate information about a product to groups and individuals. Public relations people will monitor these individuals and groups and maintain good relationship with them. They work with news and information media to ensure appropriate coverage of the product as well as the company. For publicizing product public relations people will arrange press conferences, contests, meetings, and other events that will draw attention of prospective customers to products and services.

Each of these tools and techniques of promotion mix has its own importance; hence it should be so devised as to fulfill the marketing objectives. These techniques and tools are complimentary to each other. Some such factors as nature of product, type of the market, stage of the product life cycle, budget, etc. have their impact on the promotion-mix.

4.4.1 Promotion and Library and Information Centre

Libraries and information centres generally are not directly involved in promotion activities. The use of promotional activities depends upon the type of library and its respective target users and marketing objectives. However, in the present times, they make use of some of the following methods/techniques in addition to 'word of mouth references' to promote their services and products.

4.4.1.1 Advertising

As stated above, it is considered as a non-personal promotion method of products and services of limited effectiveness. However, there are examples of libraries and information centres in India as well frequently advertising their services in newspapers with a view to bring in awareness among the target user groups.

4.4.1.2 Displays

Displays of products help in creating an atmosphere to encourage utilizing fully the resources of a library and information centre. Many libraries have floor space that could be utilized for regular displays to promote their products. Books and other information products well displayed immediately on receipt attract the attention of many a library user. Various library and information products which otherwise may remain unnoticed could prove important for some prospective users.

4.4.1.3 Personal Contact

It is regarded quite effective method for promoting library products and services. Personal contact with users with a typical “Mona Lisa smile”, being courteous and helpful goes a long way in promoting library resources and services. It may be mentioned that smile, politeness, attitude, courtesy, etc. cost nothing but give much of the needed light hence effective. The manner of library staff, in person or on telephone, will affect users’ rating of the library. Personal contact and personal skills with users lead to ‘word of mouth reference’ which proves most effective promotional method. One satisfied user can bring to the library several such new customers who might not be so responsive to other promotional methods.

4.4.1.4 E-Mail

When the library has a large number of users on e-mail it is easier way of reaching to them quickly and economically. This method of targeting users is more precise than most other techniques hence it is more effective in reaching at specific user segments.

4.4.1.5 Internet

In the emerging electronic environment, Internet has the power to improve the image of a library and information centre, and it offers the opportunity to enhance its services. It may take little more time to set up information services on the Internet, but once maintained it can help greatly in increasing levels of user satisfaction. Library and information centre may also use its Web home page as a new promotional tool advertising in-house services and/or products as well as e-information resources.

4.4.1.6 Newsletters and Leaflets

Both of them are good as well as effective means of information communication. In a library newsletter can be used to list new web sites, new journals/e-journals, online services, new databases, new products, etc. A short but crisp newsletter should be produced regularly so as to make impact on different user groups. Leaflets can be displayed on notice boards in and out side the library at some prominent place to make them more visible.

5. Seven Ps Model

The marketing mix based on the 4Ps model has been discussed above. Some authors have added one, two, or three more Ps to the existing model, making it a 7Ps model. They argue that while the 4Ps can remain the backbone elements of marketing mix, three additional elements, viz., People, Physical Evidence, and Process, are required because of more direct contact of providers and recipients, visible nature of service, etc.

Authors like Rafiq and Ahmed suggested that there is a high degree of dissatisfaction with the 4Ps model of McCarthy. Therefore, many authors extended support to Booms and Bitner's 7Ps framework. The three additional Ps are represented also by three Cs as follows:

People	--------------	Company Employees
Physical Evidence	--------------	Concrete Evidence and Experience
Process	--------------	Course of Action in Service Delivery

These are discussed below:

5.1 People

People are the most important element of any service or experience. People in an organization, working frontline as well as back stage, considerably influence customer perception of product/service and human players who take part in product/service delivery to customers. Thus, it includes organization's employees, supporting staff with whom the customers may interact. Services tend to be produced and consumed at the same moment, and aspects of the customer experience are altered to meet the individual needs of the person consuming them. It is, therefore, important that the people in the organization must be carefully selected, properly trained and motivated to deliver services to customers as they influence customer satisfaction considerably.

5.2 Physical Evidence

This is another important element in effective marketing of products and/or services. It is the material part of a service. But strictly speaking, there are no physical attributes to a service. So a customer tends to rely on certain cues. It is the environment in which the services are delivered and includes any tangible commodities that facilitate performance and communication of the services. Physical Evidence should, as far as possible, consist of the delivery of services so as to provide maximum customer satisfaction. For example, ambient conditions in a library, like a well-lighted reading area, properly stacked book shelves, spacious layout of reading area, signs, etc. would have a reasonably positive impact on the minds of library users.

5.3 Process

Process is another element of the extended marketing mix, or 7Ps. There are a number of perceptions of the concept of process. However, it includes the procedures, mechanism and flow of activities that produce the outputs in the form of products/services and deliver them. This shift from evaluating outputs only to evaluating the processes, though a recent phenomenon but organizations are slowly moving toward it.

Processes help in control of services and bring uniformity in their delivery. These procedures and activities, therefore, need to be designed carefully so as to provide quality services. For example, with the application of new technology in libraries and information centres the processes can be refined and high quality services at lower cost can be delivered to users.

6. Marketing of Information Services/Products and Library Objectives

One of the main objectives of a library and information centre is to provide information services and products to its users to meet their demands and thereby encourage the use of valuable and precious information resources. This requires a different approach, a "shift from product or service orientation to customer or need orientation."

Marketing of information products and services provides libraries and information centres with the tools for collecting and analyzing useful data about information needs of users. It is further synthesized and thereby assists in designing, developing and delivering appropriate products and services. It is argued that "needs assessment is central to any program of product development and essential to establish the targets for any marketing process." Therefore, the need to develop customer-centred and strategic market planning has now become part of effective library management.

The modern libraries and information centres can solve the problem of underutilization of their resources and services by applying principles of information marketing. The libraries and information centres can thus become market-oriented organizations and fulfill their objectives.

References

Bhat, Ishwara M. "Marketing of Library and Information Services at British Council Library Network in India". *DESIDOC Bulletin of Information Technology*, Vol. 18 (3) May 1998: 29-33.

Bryson, Jo. *Effective Library and Information Centre Management*. Bombay: Jaico Publishing House, 1996.

Cronin, Blaise. *The Marketing of Library and Information Services*. London: ASLIB, 1981.

Goi, Chai Lee. "A Review of Marketing Mix: 4Ps or More?". *International Journal of Marketing Studies*, Vol. 1(1) May 2009: 2-15.

Gupta, D.K. and Jambekar, A. "Developing a Customer-Focus Approach to Marketing of Library and Information Services". *DESIDOC Bulletin of Information Technology*, Vol. 22 (3) May 2002: 5-14.

Jain, Abhinandan, et al. *Marketing Information Products and Services: A Primer for Librarians and Information Professionals*. New Delhi: Tata McGraw-Hill, 1999.

Janakiraman, M. "Marketing Mix in the Context of Library & Information Products and Services". *DESIDOC Bulletin of Information Technology*, Vol. 18 (3) May 1998; 11-19.

Kotler, Philip. *Marketing Management: Analysis, Planning, Implementation and Control*. New York: Prentice-Hall, 1997.

Kotler, Philip and Keller, Kevin L. *Marketing Management*. 12th ed. New Delhi: Prentice-Hall of India, 2007.

Saracevic, T. "Processes and Problems in Information Consolidation". *Information Processing & Management*, Vol. 22(1) 1986: 45-60.

Saracevic, T. and Wood, J.B. *Consolidation of Information: A Handbook on Evaluation, Restructuring and Repackaging of Scientific and Technical Information*. Paris: UNESCO, 1981.

Seetharama, S. *Information Consolidation and Repackaging: Framework, Methodology, Planning*. New Delhi: Ess Ess Publications, 1997.

Information Products—Handbooks and Newsletters

9

1. Introduction

The methodology for the preparation of information consolidation products has been discussed in detail in earlier chapters. Information products are prepared by selecting, acquiring, evaluating, analyzing, synthesizing, restructuring, packaging, repackaging information gathered carefully from various sources. These information (consolidation) products include handbooks, manuals, state-of-the-art reports, reviews, trend reports, guidebooks, do-it-yourself booklets, and so on. These publications provide information on various attributes of the product, process, material, service, and also promote their use as well as sale. Recent advances in computers and telecommunication technology have great impact on communication pattern of industrial, commercial and public service organizations including libraries and information centres.

2. Handbooks

A Handbook is defined as a comprehensive and detailed work on a particular topic for practitioners, structured for quick reference. It is called as a handbook because it is usually convenient to be held in the hands of the users. It is, therefore, a ready reference book the relevant portions of which are consulted at the time of need. A handbook is the best 'working tool' of a technologist. Handbooks usually contain factual information, data, tables, formulae, etc.

A Handbook is a type of reference source, or other collection or compendium or instructions, that is intended to provide ready reference. The term handbook is commonly used interchangeably with a Manual because, according to William Katz, it is difficult to distinguish between the average handbook and the average manual. He stated that a handbook "is a collection of a miscellaneous group of facts centred on one theme or subject area."

Louis Shores defines a handbook as "A reference book of miscellaneous facts and figures on one or many subjects assembled for ready use, in response to popular interest or to a specific need for concise, handy information."

A great number of organizations, centres, and other institutions have been preparing handbooks since a long time. The term handbook is sometimes used for documents that are produced within an organization. A handbook may deal with any topic, and is generally a compendium of information in a particular field or about a particular technique. Such handbooks are designed in such a way that they may be easily consulted to obtain quick answers in a certain subject, i.e., they are designed to meet what is called 'everyday approach'.

Professional organizations usually prepare a variety of handbooks ranging from short duplicated set of regulations to glossy multi-coloured reports. These are produced for different type of users such as technical personnel, industrial workers, office staff, and so on. On the other hand, some handbooks are prepared for general information, such as history of an organization; and there are other handbooks that are meant only for the benefit of the employees of the organization, for example, they provide a set of rules; and so on.

Some handbooks are not necessarily pocket-sized to be held in hands, but they do provide ready reference, as is the case with several engineering handbooks. For providing ease of consultation most of them are single-volumed handy compilations.

3. Purpose

Katz states that the primary purpose of handbooks is to serve as ready-reference source for given field of knowledge. Handbooks would normally lay emphasis on established knowledge rather than on recent advances, although in the field of science, handbooks that are more than a few years old may not be of much use. They primarily give information about 'how to do', 'how to perform', 'how to make', etc., i.e., these consist of 'facts to know' and 'instructions to do'.

4. Scope and Coverage

Most handbooks have limited scope, and may include the following in their scope and coverage:

(a) provide facts, figures and themes of a specific area.

(b) no emphasis on newer ideas and current developments.

(c) help in understanding the subject.

(d) prepared with the assumption that users have some background knowledge about the subject covered.

(e) provide sources for the facts and figures included.

(f) gives instructions on how to serve as guides to occupations, hobbies, art forms, trades, etc.

5. Types of Handbooks

Handbooks may be: General handbooks, Historical handbooks, Literary handbooks, Statistical handbooks, Official handbooks, Practical handbooks, Subject handbooks, etc.

According to Saracevic and Wood, there are a number of types of handbooks; the main ones are as follows:

5.1 Compilation of Critical Data and Formulae

This type of handbooks is usually on a substance, process, observation and/or phenomenon. These are used mostly in science and technology for experimentation, design calculations, construction, safety considerations, maintenance, replacement, comparison, etc. Similarly, publications of evaluated data from social science data archives are also examples of this type of handbooks.

5.2 Guidelines for Accepted and Tested Procedures

These handbooks are aimed at actually doing things, for decisions related to processes and procedures, and for learning as well. They range from step-by-step pamphlets, to how-to-do books and manuals, to extended discussions of options in given processes.

5.3 Description and/or References of Standards

This type of handbooks generally provides descriptions or references of various standards that can be applied in doing things. These handbooks can incorporate required standards (as in electrical installations) or recommended standards (as in bibliographic work). They can range from tables, to descriptions, to discussions aimed at promoting harmonization and interconnections.

6. Methodology for Preparation of Handbook

While preparing a handbook it is important to know about the objective and the needs of the users. It must be kept in mind that the style of writing and presentation has to be simple, easy to read and understand. As far as possible, complex sentences and technical jargon should be avoided.

For preparing a handbook, the following sequence of steps is generally followed.

6.1 Collection of Material

The first activity to be pursued for this purpose is to collect material from different information sources for inclusion in the handbook. The knowledge and experience of the person who is assigned the job of writing the handbook is significant as he may be a subject specialist.

6.1.1 Sources of Information

The writer and his team have to know about the various types of documentary, human, and institutional sources of information. This would facilitate the location of the relevant sources. In case a published record is not readily available, raw data may be obtained from the concerned information generating agency. The three major categories of sources are as mentioned below:

(i) ***Institutional Sources:*** The main institutional sources of information include in the three major divisions: (a) international, (b) national, and (c) local institutions.

- Central and sate governments and their various departments,
- Learned and research institutions at various levels,
- International organizations,
- Professional associations,
- Financial institutions,
- Educational institutions, and so on.

While preparing the handbook, take note of those institutional sources of information that are on the subject concerned. It will help to identify their conferences, projects, specializing in the subject by way of research, education, etc. Thus, these institutions will be able to provide information on many aspects of the subject which is not readily available from other sources.

(ii) ***Documentary Sources:*** The writer and his team will take care of the various documentary sources of information on the subject of the handbook. The main categories of these sources include: (a) primary, (b) secondary, and (c) tertiary sources. Some of these are listed below:

- Research reports
- Research and technical periodicals of primary nature
- Conference, seminar, and workshop proceedings
- Advanced treatises
- Descriptive sources, like economic surveys, plan documents, etc.
- Trend reports
- Newsletters
- Books
- Letters

- Notes in files
- Statistical sources
- Geographical sources
- Regulatory sources, like gazettes, codes, handbooks, etc.
- Bibliographies, including ad-hoc bibliographies
- Reference sources, like encyclopedias, directories and yearbooks, etc.
- Indexing and abstracting periodicals
- Review documents (annual reviews, advances in, year's work, etc.)
- List of organizations specializing in the publication of documents in the subject concerned.

(iii) Human Resources: Adequate knowledge about the human resources of information on the subject concerned is very helpful in identifying the living persons recognized as specialists. These subject specialists, when consulted may be in a position to furnish some useful information on the subject which is not readily available from other sources. These specialists ought to be of reputation, as far as possible, at the local, national, and international levels.

6.2 Information Analysis

After collecting the material and other requisite information from the above and other sources, it is analyzed so as to arrange it in chapters. The analysis is done to ensure that the chapters have comprehensive coverage of information as per users' requirements. It also helps in evolving an optimal mix of text, tables, illustrations, etc. for each chapter of the handbook.

6.3 Structuring of Contents

The structuring of contents is done to assure optimum coverage of each chapter of handbook. The contents are determined by the purpose and readership of the handbook under preparation. When the purpose is to inform new employees of the fundamental rules and working conditions of the organization, the handbook may contain the following aspects:

- – Organization of material—chapters, general subjects, etc.
- – Aids to users—table of contents, comprehensive subject index
- – Text and tables, or tables or illustrations only
- – Instructions and directions
- – Scope—subject covered
- – Period covered, revisions, if any
- – Illustrative materials, charts, maps, pictures, etc.

Thus, a document of balanced coverage, keeping in view the information needs of the users, may be prepared and published.

6.4 Parts of Handbook

A handbook may consist of the following parts:

(i) **Title Page and Cover:** It will include title of handbook, name of organization and/or individual/s preparing it, month and year of publication.

(ii) **Table of Contents:** It consists of numbers and titles of chapters and/or sections, and the page number on which each item begins.

(iii) **List of Tables:** If the number of tables provided in the handbook is more than five, a list containing the table number, title, and page number, has to be provided in the beginning.

(iv) **List of Illustrations:** In case a large number of illustrations is used in the text, a list may be provided with title and page number of each illustration.

(v) **Introduction:** An introduction usually indicates the purpose of preparing the handbook. It defines the scope of the subject covered, its historical perspective, etc. of the handbook. Further, it may also include an overview of the topics covered, constraints in preparation, etc. It also contains a brief guide to use the handbook, and the meaning of any abbreviations, symbols, or technical terms used in the text.

(vi) **Preface:** It is usually provided by the Director or Head of the institution concerned. It indicates general reference to the efforts made in its preparation by the individual/s, and the value of such a work to the end users.

(vii) **Body:** In this part of the document the subject matter appears in a classified order including logical arrangement of various parts, sections and sub-sections, and chapters depending upon its nature and size. The body of the handbook forms of details about historical perspective, scope and coverage, methodology, literature review/literature consulted, presentation of data in textual and tabular form, results, findings, and conclusions.

(viii) **Bibliographical References:** Published and/or unpublished sources of information used for preparing the handbook should be cited in the text and a listing of all the references has to be given at the end of the body. The citations should be provided with complete bibliographic details, with reference numbers, and in proper order as per a standard style manual. Sometimes, the references can also be given in the form of a bibliography.

(ix) **Appendices:** At the end of the body of the text, or the chapters, some information worth recording and likely to be useful to the users but not included in the text/body, is provided in the Appendix. This is done to avoid the details about some issues and keep the handbook easy to read and understand.

(x) **Glossary:** It is a list of technical terms used in the handbook. If the number of such terms is small, they may be explained in the footnotes on the particular page they occur. But in case the list is long, each term may be provided with a brief definition in the end.

(xi) **Index:** In case the handbook is large in size and the table of contents is not adequate to help the users to locate a topic, or subtopic, etc., a subject index must be provided. The entries in the index are arranged in alphabetical order and are cross referenced. The page number/s is/are mentioned against each entry in the index.

6.5 Scrutiny

Seetharama suggests that the manuscript/typescript prepared by the author/s on the basis of the inputs as described above, may be scrutinized by an expert committee which can make suitable textual and language corrections.

6.6 Editing

The arrangement of various elements of the structure of handbook should be according to some standard style manual. Technical editing of the typescript is therefore necessary to assure the correctness of the contents. On the other hand, language editing is done to maintain a uniform style of presentation and for the ease of understanding of the book by using a simple language. Unnecessary use of flowery language and jargons should be avoided, and due care should be given to clarity of language.

6.7 Compilation

The final methodical compilation of chapters, tables, illustrations, may be done under the supervision of a committee set up earlier for the purpose of preparation of the handbook.

6.8 Production

Before going for the actual process of production, decisions about the size of the handbook, number of copies to be printed, and the mode of production, etc. have to be taken by the concerned committee. It could be got printed and published by one of the leading publishers so that not only the cost of production is borne by it but the responsibility of marketing also rests with the publisher.

Examples of Handbooks

These days almost every conceivable subject produces handbooks at one or the other time. Some helpful examples are given below:

1. Saracevic, Tefco and Wood, Judith B. *Consolidation of Information: A Handbook on Evaluation, Restructuring and Repackaging of Scientific and Technical Information.* General Information Program and UNISIST of the UN Educational, Scientific and Cultural Organization, Paris, 1981 (PGI-81/WS/16).
2. *MLA Handbook for Writers of Research Papers.* Modern Language Association of America (it is based on the MLA Style Manual, and serves as an academic style guide widely used all over the world).
3. *Handbook of Chemistry and Physics.* Boca Raton, FL: Chemical Rubber Company Press, 1913 to date, Annual.
4. *The New Guinness Book of Records.* London: Guinness World Records Ltd, 1955 to date, Annual.

MANUALS

1. Introduction

A manual is a concise reference book dealing with a particular process, procedure, operation, mechanism, job instruction, etc. It is essentially a how-to-do-it document prepared with meticulous care containing details of the concerned subject matter. It explains how a certain thing happens, or how a particular job should be done. However, sometimes it is difficult to distinguish it from an average handbook.

A manual grows out of sheer necessity and the concern of an organization is to keep running smoothly and getting jobs done efficiently.

2. Scope

A manual, like a handbook, has a limited scope, and it usually focuses on a specific area of interest. According to

Katz, its particular value is its depth of information in a narrow field of knowledge.

3. Uses of Manual

Some of the uses of a typical manual are as under:

(i) Provides instructions to users on the uses of equipments/appliances/software, etc.

(ii) Provides instructions to users' representatives on how to instruct their own employees about the use of various machines and other equipments.

(iii) Provides instructions to management, supporting staff or helping staff on the use of equipments and/or procedures, etc.

(iv) Provides instructions to other departments or organizations on how to liaise their operations with yours.

4. Types of Manuals

Different types of manuals dealing with the description of various jobs, functions, instructions, procedures, software, etc. are available in the present day professional world. Accordingly, they can be grouped in the following main categories:

4.1 Technical Manuals

A technical manual describes an equipment, or a process, or a software, or a mechanical device and contains instructions for installation, operation, maintenance, repair, etc.

4.2 Procedure Manuals

A procedure manual deals with established procedures and working practices in an organization. This type of manuals is helpful for the managers in carrying out their work in methodical manner, therefore, mostly they are non-technical.

4.3 Laboratory Manuals

The category of laboratory manuals describes how the users can perform experiments and give information about the concerned equipment, procedure, installing software, testing techniques, etc. These manuals are prepared for specific purpose and usually accompany some equipment, process, or software.

Examples

1. *Manual of Local Studies Librarianship,* Edited by M.A. Dewe, 2 Vols. Aldershot: Gower, 1987 and 1991.
2. *The Merck Manual of Diagnosis and Therapy.* Rahway, J.J.: Merck and Company, 1899 to date.
3. *Haynes Owner's Workshop Manuals.* England: Haynes Publishing Group, 1960 to date.

NEWSLETTERS

1. Introduction

Newsletters are new information product in the modern information world. They have been with us for more than 500 years in one form or the other as the earlier newsletters or newssheets were published in Europe's commercial centres. These newsletters contained business news and news of other public events likely to interest many sections of the society.

Thus, newsletters are forerunners of modern newspapers which included news of public interest and meant for general circulation. However, there has been rapid growth of newsletters during the twentieth century.

Newsletters are an important source of very valuable information. According to Christine D. Reid, they can provide timely, very focused and detailed information frequently not found anywhere else. They can also provide a specialized slant on the news.

2. Definition

S. Seetharama defines a Newsletter "as a serial consisting of one or few pages, usually printed, containing nascent information or news which is of interest to a special or particular group of users and which is published periodically."

Newsletters are, therefore, periodical publications brought out by various organizations, institutions, banks, companies, government departments, professional societies, and trade and industry associations for providing speedy information to their members and/or subscribers. These are also issued by commercial and research organizations to make their users aware of the new products that are being brought out in the market. Some organizations publish newsletters as a substitute to journals, and contain limited pages supplemented with a few photographs as well.

3. Types of Newsletters

Based on the publishing organizations, newsletters may be categorized as follows:

(i) R&D Organizations
(ii) Associations
(iii) Government Organizations
(iv) Public Sector Undertakings
(v) Private Institutions
(vi) Academic Institutions
(vii) International Organizations
(viii) Autonomous Institutions, etc.

4. Objectives

Newsletters are regarded as simple publications conveying up-to-date information usually relating to a specific society or business organization. These are easy, fast, and relatively inexpensive to produce. They are prepared to fulfill the following objectives:

(a) to publicize and promote products and services of the organizations and institutions;
(b) to announce current and forthcoming events like, holding of conferences, seminars, workshops, etc.;
(c) to communicate information about social and cultural activities, appointments, promotions, transfers, retirements, obituaries, etc. of the members/subscribers of the concerned organizations;
(d) to publish timely and relevant activities of current interest.

5. Contents

Based upon the purpose, different newsletters cover different type of information. For example:

5.1 R&D Organizations

In their newsletters, they include information about programmes of the organization, grants and fellowships, awards and rewards, forthcoming conferences/seminars/ workshops, etc.

5.2 Associations

Their newsletters publish information about President's or Secretary's report, announcements of association, employment news, personnel information, academic and/or professional achievements, and other news of interest to their members.

5.3 Industrial Organizations

They publish newsletters for (a) internal circulation, and (b) external circulation. The one for internal circulation has news about and for the employees such as promotions, transfers, announcements of welfare measures for the employees, social and cultural activities of the employees, etc. On the other hand, the one meant for external circulation covers information about new products and services, highlights the achievements of the organization, announces benefits and discounts for the customers, etc.

6. Frequency

The frequency of publication of newsletters varies according to their purposes, nature of information covered and the ambit of organizations. Some newsletters are published every week, while others are brought out fortnightly. There are some others which are published monthly or bi-monthly, or quarterly, half-yearly, or even yearly.

7. Methodology for Preparation of Newsletters

The steps involved in the preparation of good quality Newsletters conform, generally, to the methodology for the preparation of Information Consolidation Products discussed in Chapter 7. However, some relevant information is given here.

7.1 Identification of Users and Their Information Requirements

In view of the institutional requirements for information on projects, etc., individual user interests, it is necessary to conduct a user survey and study.

7.2 Scope Determination

The subject scope of Newsletter to be produced is determined by expressing the totality of the subject coextensively and uniquely.

7.3 Collection of Information

For preparation of Newsletter the sources of information may be:

(i) **Documentary:** These may be internal or external to the organization.

(a) **Internal Sources:** These may include:

- the research projects in progress or completed in the organization
- information gatekeepers
- information services rendered by the information centre

(b) **External Sources:** These may include:

- Periodicals, technical and research reports, conference/seminar proceedings, patents, standards, specifications, trade catalogues, government and non-government publications, project reports, reviews, preprint, data sheets, drawings, newspapers, photographs, etc.

(ii) **Non-documentary**

Some of the non-documentary sources may include the following:

- Information centres, Institutions, Data banks, Clearing house, etc.

– Vendors, Suppliers, etc.

– Consultants, Guides, etc.

From the above and other sources of information, the information professional should scan only the current and relevant information to the users concerned, for presentation in Newsletter.

7.4 Appraisal of Information

The information collected from different sources, for inclusion in the Newsletter, is further critically examined from the point of view of its relevance, currency, accuracy, adequacy, etc. Appraisal of information should be done in view of the scope determined for this purpose. For example, Seetharama observes that the academicians and researchers would prefer information on current and forthcoming activities which are academic and research oriented; and so on.

7.5 Presentation of Information

It involves the following two processes:

(i) Choice of textual and/or non-textual mode of presentation.

(ii) Repackaging of information to suit with users' requirements.

However, to make it easy to use, a Newsletter may adopt an optimal mix of text, tables, illustrations, etc. Technical terms may be used to a limited extent so that these could communicate the ideas better. Adequate, appropriate, and eye-catching featuring should be adopted for a Newsletter.

Timeliness in publication of a Newsletter is very important as it strengthens the confidence of its users.

8. Electronic Newsletters

Recent advances in computer and telecommunication technology have led to the development and application of computer networks. These, in turn, have helped in the development of electronic newsletters which are being used by the users since 1990s.

Electronic newsletters are distributed by e-mail to those users who would willingly subscribe to them online. This method of sending the e-Newsletter is regarded as fast and cost effective. The e-Newsletters have gained rapid acceptance for the same reasons email in general has gained popularity over printed correspondence. The use of e-Newsletters is becoming quite popular for advertising the products and services in business and industry because no printing and mailing costs are involved.

One can start with *Predicasts Newsletters*, available online from DIALOG, DataStar, etc. providing access to the full text of more than 600 specialist newsletters.

Similarly, *NewsNet* is an online system based in Pennsylvania and devoted entirely to newsletter coverage. It carries more than 800 full-text newsletters.

In the present environment, more and more sources are becoming available electronically to a large number of users.

8.1 E-Newsletters: Advantages

Some of the advantages of e-Newsletters are identified as follows:

(i) These are convenient and delivered to the users at their information desk without delay.

(ii) E-mail newsletters have timely information and real-time delivery.

(iii) These are informative and keep users up-to-date.

(iv) Sometimes users forward e-Newsletters, if found useful, to friends and/or colleagues.

CompanyNewsletter.com Inc., Lakeville, USA has been producing high quality custom newsletters for businesses since 1996.

Examples

1. *CSIR News*. It is a fortnightly newsletter published by the National Institute of Science Communication and Information Resources (NISCAIR), CSIR, New Delhi.
2. *Financial Times Business Reports*. These provide current market and industry news covering the areas

of business and finance, energy, media and technology. These are accessible online via FT PROFILE and DataStar.

3. *Predicasts Newsletters*. Available online from FT PROFILE, DIALOG and DataStar. This provides access to the full text of more than 600 specialist newsletters.
4. *NewsNet*. It is an online system devoted entirely to newsletter coverage. It carries over 800 full-text newletters.

HOUSE JOURNALS

1. Introduction

House journals are believed to have a distant past when they were used for internal communication. Later, the term "house organ" was also used for such publications. In recent times, however, the area of work of house journals has expanded many times in many directions: they are increasing circulation, adding more pages, and publishing more and more new house journals to meet the demand of special interest groups.

A house journal is defined as an internal publication produced by an organization in order to inform and entertain its employees and to generate better internal communication and relationships.

House journals are serial publications, like the newsletters, published by industrial and commercial organizations, corporations, business houses, public sector undertakings, and other similar organizations for circulation among their customers, shareholders, employees, dealers, retailers, etc. to inform them of their products, performance and other activities.

2. Objectives/Functions

Some of the objectives and functions of house journals are as follows:

(i) to promote and enhance the image of the organization as well as that of management among its customers and employees.

(ii) to improve business and industrial relations.

iii) to keep employees and customers in the picture.

(iv) to provide an objective news service.

3. Types of House Journals

All house journals can be categorized into the following two distinct types:

3.1 Internal House Journals

These are published for employees and members of the organization. These are primarily meant for internal circulation, and they inform about the welfare measures taken by the management for the employees.

3.2 External House Journals

An external house journal could be in the form of a standard technical journal, or a magazine, or a periodical catalogue. These house journals are brought out for customers or potential customers of the organization and are meant for external circulation.

3.3 Combination

Sometimes house journals may be published as a combination of both internal and external wherein attempt is made to include features of both types of house journals.

4. Contents

Although the contents of house journal vary with the type of house journals, yet some standard information could be as follows:

(i) A message from the Head of the Institution/ organization is a regular feature, usually on the topic of current interest.

(ii) A brief editorial from the Editor of the house journal.

(iii) Information about the cultural activities related to the organization, like sports events, annual day, etc.

(iv) Information about personnel changes, incentives, awards, rewards, bonuses, promotions, transfers, retirements, etc.

(v) Articles, features, stories, etc. by employees.

5. Characteristics of House Journals

Some of the common characteristics of house journals are as follows:

(i) Produced by organizations for enhancing their image and promoting the products and services.

(ii) Distributed free to the employees and potential customers.

(iii) Printed on good quality paper and are usually made attractive.

(iv) Usually carry no advertisements.

6. Methodology for Preparation of House Journals

The methodology and guidelines for preparation of house journals are the same as that of the Newsletters.

Examples

1. *Image*. A monthly house journal published by Indian Airlines Ltd., Mumbai.
2. *DRDO Newsletter*. A monthly house bulletin published by DRDO, DESIDOC, New Delhi.

References

Katz, William. *Introduction to Reference Work*. 8th Ed. New York: McGraw-Hill, 2001. 2 Volumes.

Lea, Peter W. and Day, Alan, Eds. *The Reference Sources Handbook*. 4th Ed. London: Library Association Publishing, 1996.

Saracevic, T. "Processes and Problems in Information Consolidation". *Information Processing and Management*, Vol. 22(1) 1981: 45-60.

——. *A Course in Information Consolidation: A Handbook for Education and Training in Analysis, Synthesis and Repackaging of Information*. Paris: UNESCO, 1986.

Saracevic, T. and Wood, J.B. *Consolidation of Information: A Handbook on Evaluation, Restructuring and Repackaging of Scientific and Technical Information*. Paris: UNESCO, 1981.

Seetharama, S. *Information Consolidation and Repackaging: Framework, Methodology, Planning*. New Delhi: Ess Ess Publications, 1997.

Wetherbe, J.C. *System Analysis and Design*. 3rd ed. St Paul, USA: West Publishing Company, 1988.

Reviews and State-of-the-Art Reports

10

1. Introduction

In the modern society there is influx of ideas and deluge of information. Many fields of human, scientific, technical, commercial, health, social, and cultural endeavours have increasingly become involved with information. Information, therefore, has become a vital resource, and has assumed so much significance as to be required for socio-economic development of any country.

Information, produced by subject specialists is not easy to understand by an ordinary user because it is not written in a simple language. Again, access to this information is not easy for a potential user as there is too much information available on a topic of his interest; hence he is unable to use it.

Attempts have been made to find solution to such problems coming in the way of access and use of valuable information. For instance, solution is found in bibliographic control and organization along with indexing and abstracting of information. But many information related problems still remained unaddressed. Therefore, it was recognized for a long time that other solutions are required to overcome this difficulty.

The users require information packaged in a language, form and manner which they not only can understand easily but use it also to meet their needs and demands. Information consolidated and repackaged is considered more appropriate to users at different levels, capacities, assimilation, etc. It means providing right information to the right user in the right form, and at the right time.

Many information consolidated products are now being designed and developed for large segments of users. Some of them have been discussed above; and some more like reviews, state-of-the-art reports, etc. will be discussed here.

2. Reviews

2.1 Definition

Indexing and abstracting services, among others, have been useful tools in bibliographic control of information. Meanwhile, the number of indexing and abstracting services has also gone up tremendously making it difficult for individual users to have a systematic access to recently published literature.

It may be pertinent to mention that 'review' here is different from a book review wherein is provided only an assessment of the publication; it is also different from a summary or bibliographic essay; it is rather a critical and evaluative review of literature in a particular subject.

In order to have a coherent view of the overall development of a subject, such information products like reviews have been prepared. According to Guha a review is a "narrative account of the progress of a particular field of study, prepared by an expert in the field". Saracevic and Wood have defined a review as "a critical synthesis of the state of knowledge in a given subject or topic; it is a critical examination of information and literature on a subject or topic accommodated in its broader framework."

Reviews are prepared by information selection, analysis, evaluation, restructuring and repackaging for satisfying information requirements of users working in a given field.

2.2 Characteristics

Reviews have the following characteristics:

1. Integration: A review arranges each of the articles reviewed in a context and integrates them with the existing body of knowledge on that subject.

2. Evaluation: Critical reviews are invariably evaluative in nature. While evaluating, the review sifts through the available literature, identifies significant items and in the process eliminates some of the primary documents considered not useful.

3. Compaction: It is an important characteristic of reviews. Compaction is usually estimated by determining (i) the average number of references cited per page of text in the review, or (ii) the ratio of the number of pages of primary documents to the number of pages in the review document. However, the degree of compaction varies from subject to subject.

2.3 Types of Reviews

There are two types of reviews as given below:

1. **Bibliographic Reviews:** This category of reviews is literature-oriented or bibliographic approach, wherein it selects and evaluates primary literature published on a subject within a defined period of time, say annually, biannually, or five yearly, etc.
2. **Instructional Reviews:** This type of reviews follows subject-oriented or didactic approach. It focuses on a particular scientific or technical problem, and attempts to offer its solution.
3. There are some other reviews which combine the bibliographic and instructional role into one. In these reviews proportion of bibliographic or instructional type varies according to the aim and users for the reviews.

2.4 Functions

According to A.M. Woodward reviews are prepared to perform the following functions:

2.4.1 Historical Functions

These are related to the development of a subject or topic, and are as follows:

1. Evaluation of the published literature on a subject by expert and selecting significant items only.
2. Collation of information from various sources and putting them in proper perspective.
3. Compaction of existing knowledge, i.e., having new information and leaving the other known details.
4. Replacement of many scattered documents with one.
5. Identification of emerging discipline through reviews.
6. Direction for further research, suggesting new areas.

2.4.2 Contemporary Functions

These functions are user related hence beneficial to individual user. Some of the important contemporary functions are as follows:

1. **Current Awareness Function:** It keeps the individual user up to date about the current developments in the subject field.
2. **Informed Notification Function:** Most significant contributions, with full bibliographical details, are brought to the notice of the individuals through reviews.
3. **Back up for other Literature Searching:** Reviews with extensive bibliography are the starting point for more current literature yet to be covered.
4. **Tutorial Function:** Reviews are useful to students in understanding new subject, and to specialists in continuing education and obtaining overview of the subject.
5. **Inspiration Function:** Good and critical reviews provide ideas and suggestions for further research work in a subject field. It stimulates for putting research into practice.

6. **Feedback Function:** Reviews provide feedback to a researcher about his/her own published work if that is included in the review on the subject.
7. **Promote Cross Fertilization of Ideas:** Reviews in a specialist's related areas of research help in cross fertilization of ideas and promote research in new areas.

According to Saracevic and Wood, the patterns of use of reviews for individual functions may differ significantly between various types of users according to their subject interests, place and type of work, etc. Reviews can also be specifically structured and oriented to emphasize some of these functions and encourage some of the uses.

Examples

1. *Annual Review of Information Science and Technology*. American Society for Information Science, USA, Annual.
2. *Advances in Solid State Physics*.
3. Kunz, Werner; Rittel, H.W. and Schwuchow, Werner. *Methods of Analysis and Evaluation of Information Needs: A Critical Review*. Munchen: Verlag Dockumentation, 1977.

3. State-of-the-Art Reports

3.1 Introduction

State-of-the-art reports are another important information consolidation products required by scientists and other researchers. These are the type of reviews which do not have an all encompassing scope and historical orientation. These information products lay emphasis on recency and up-to-dateness of a situation.

According to Saracevic and Wood: "State-of-the-art report is a review concentrating on the most up to date information and literature in a given subject or topic".

State-of-the-art report is the highest level of development, as of a device, technique, or scientific field, achieved at a particular time.

3.2 Aim

The basic aim of state-of-the-art reports is to describe a very recent situation when they reach their users. Thus, they are timelier than the traditional review, so they are mostly a current awareness tool. In order to achieve this currency, state-of-the-art reports, observed Saracevic and Wood, are often published as informal reports, prepared on demand, restricted to a specific segment of users, sold at high prices, and becoming obsolete quite fast.

3.3 Characteristics

Some of the general features of state-of-the-art reports are as follows:

1. **Completeness:** It indicates the degree to which information and literature on a subject are covered.
2. **Perspective:** It involves the direction, purpose, subject orientation, degree of appropriateness in relation to both a given subject and for a given user group.
3. **Analysis:** It is with regard to thoroughness, depth and extent of analysis.
4. **Synthesis:** It is the degree of compactness and consolidation from various collected sources of information.
5. **Value Addition:** It is related to the requirements of the users which in this case are the subject specialists.
6. **Utility:** It is the degree to which the state-of-the-art report serves multiple functions.

3.4 Types

Saracevic and Wood stated that the state-of-the-art report is used mostly in relation to technological topics, and in business and commerce. A few major types of state-of-the-art reports and their uses are given below:

3.4.1 State-of-the-art of a Technology

These reports summarize, compare and evaluate the advances, characteristics and/or utilization of a given technology or technological product or process. The scope and emphasis differ according to intended use and users. These reports generally highlight the following aspects of a technology:

(i) **Technical and Engineering Aspects:** It indicates that these reports are oriented toward technical personnel and engineers in particular. These provide technical details of the new technology. Such reports are useful in technology design, for making purchase recommendations, for comparisons of characteristics with another alternative technology, and for keeping up to date with technology.

(ii) **Use Aspects:** These are oriented towards users of a technology. These reports emphasize on characteristics, requirements and economics of use. These reports cover aspects such as comparative reliability, durability, complexity, operating conditions, etc. of a given technology.

(iii) **Management Aspects:** These are oriented towards managers, administrators, and economists. Such reports highlight the requirements and comparisons in terms of resources (economic, physical, human, raw materials, etc.), planning, alternative choices, negotiating (licensing, purchase, and exchange), related experiences in other industries, organizations, or similar situations, synthesizing drawbacks, availability, etc.

(iv) **Strategic Aspects:** These are oriented towards policy makers on a higher level, such as lawmakers, ministry officials, government representatives, planners, officials from funding agencies, etc. Such reports highlight the impact (social, environmental, economic, etc.), strategic requirements, political considerations, international aspects, experiences of other countries, particularly in similar situations, etc.

3.4.2 Market Reports

Saracevic and Wood held the opinion that these reports usually summarize the state of an industry or a market in terms of its existence, financial strength, economics, profitability, deployment, growth, characteristics, gaps, trends, potential, etc. Like technological state-of-the-art reports, these reports can also be oriented toward different uses and clientele. Their main use is in market decisions and planning, considerations of competition, opportunities, and the like.

3.4.3 Statistical Reports

These reviews present statistical correlations in a wide variety of technical, market, demographic, scientific and other areas. Such reviews are specifically composed and evaluated to show cross-dependence of trends, situations, and results. These may accompany any of the above-mentioned state-of-the-art reports or issued as independent publications. These statistical reports are usually used for briefing decision makers.

3.5 Functions

Some of the functions of the state-of-the-art reports as stated by Woodward are as follows:

3.5.1 Peer Evaluation of Published Literature

It involves getting the published literature evaluated by peer experts in a given subject field. This brings quality information and literature to the notice of the users for further use.

3.5.2 Collation of Information from Different Sources

This acts as a unifying process for a subject or topic and provides a continuing conceptual framework in which individual works may be viewed in perspective.

3.5.3 Compaction of Existing Knowledge

It is extracting the parts of the papers containing new information only, leaving out portions that pertain to verifications, discussions, methods, etc. already known.

3.5.4 Replacement of Primary Documents as the Written Record

This involves replacing of many scattered documents with only one document.

3.5.5 Identification of Emerging Specialties

This is carried out by bringing together various works; the emergence of a new area may well be first identified in a review.

3.5.6 Direction of Research and Work

Various suggestions for further research work to be carried out are put forward.

3.6 Methodology for Preparation of State-of-the-Art Report

The methodology and guidelines for the preparation of a state-of-the-art report are the same as for the preparation of any other information consolidation product (ICP). It involves the various operations as in the case of any other ICP. These are summarized as follows:

3.6.1 Survey of Potential Users

It is an important step to study the information needs of potential users of a state-of-the-art report. It involves knowing the specific information about users like their subject and related interests, etc. This could be done through questionnaires, interviews, personal contacts with users, feedback from users, etc.

3.6.2 Collection of Information

It involves identification and selection of sources potentially containing the most useful information for a given user problem and information needs. This selection is usually done from a variety of primary and secondary sources of information on the subject under consideration.

3.6.3 Evaluation of Selected Information

Evaluation is the process of ascertaining and appraising the relevance, intrinsic merit, validity and reliability of the information selected. It also involves comparison of information on the same subject or topic from a variety of sources. In case of any ambiguity, the original documents may also be checked.

3.6.4 Analysis of Information

In this step analysis of information is done in order to determining and isolating the most salient and significant information as conveyed by various information sources.

3.6.5 Merging of Information

It is the process of combining the most significant information as analyzed from a variety of information sources collected for the purpose.

3.6.6 Restructuring of Information

This step involves the restructuring of extracted information into a content that can be used most effectively and efficiently by the target segment of users. Restructuring of information involves presentation of analyzed information as a new whole which differs from the original arrangement. It also involves synthesis, condensation, rewriting, simplifying, and review of information. It can be decided at this stage in which form information can be presented, that is, as a review, or state-of-the-art report, or any other form. Simplifying and rewriting of information concerns with regard to the expression of information which may originally be beyond comprehension, in a technical language, to a language and form that will be within the understanding level of the users.

3.6.7 Packaging

Packaging of information concerns with the physical recording or arrangement or presentation of information in a given medium. The restructured information is now packaged in a format or medium that will enhance the potential of the use of the product by its users.

3.6.8 Repackaging

Repackaging of information deals with rearrangement of physical format or medium tailored to the requirements of a specific segment of users. In other words, repackaging of information refers to the presentation of information in more understandable, readable, acceptable, and usable forms. Thus, it helps as a means to prompt delivery of relevant information.

3.6.9 Dissemination of Information

Dissemination of information is carried out in such ways as will encourage and promote the use of the state-of-the-art report. Dissemination is a process of conveying information through given channels to users. It includes the spreading about, distribution and delivery of information product and/ or service.

3.6.10 Feedback

It involves feedback from the target group of users. It will throw light on the lacunae and shortcomings, if any, existing in the state-of-the-art report. On the basis of the feedback received, evaluation of the product can be done and necessary corrections may be initiated so as to present the final product in a much better way.

The above-mentioned guidelines and/or steps will go a long way in helping the preparation of a state-of-the-art report, or for that matter any other information consolidation product.

Examples

1. *Information Technologies: State-of-the-Art and Imperatives for India*, Vol. II. New Delhi: TIFAC, Department of Science and Technology, 1993.
2. *State-of-the-Art Report on Technology for the Efficient Management of Geo-spatial 3D data on Commodity and Distributed Platform.* EU Project. V-PLANET (IST-2000-28095). March 2002.
3. *Advances in Librarianship.* Annual. New York: Academic Press, 1977.
4. *Progress in Operations Research.* Annual. New York: Academic Press, 1969.

References

Katz, William A. *Introduction to Reference Work.* 8th ed. New York: McGraw-Hill, 2001. 2 Volumes.

Lea, Peter W. and Day, Alan, Eds. *The Reference Sources Handbook.* 4th ed., London: Library Association Publishing, 1996.

Saracevic, T. *A Course in Information Consolidation: A Handbook for Education and Training in Analysis, Synthesis and Repackaging of Information.* Paris: UNESCO, 1986.

——. "Processes and Problems in Information Consolidation". *Information Processing and Management*, Vol. 22(1) 1981: 45-60.

Saracevic, T. and Wood, J.B. *Consolidation of Information: A Handbook on Evaluation, Restructuring and Repackaging of Scientific and Technical Information*. Paris: UNESCO, 1981.

Seetharama, S. *Information Consolidation and Repackaging: Framework, Methodology, Planning*. New Delhi: Ess Ess Publications, 1997.

Wetherbe, J.C. *System Analysis and Design*. 3rd ed. St Paul, USA: West Publishing Company, 1988.

Woodward, A.M. "The Roles of Reviews in Information Transfer". *Journal of the American Society for Information Science*, Vol. 28(3) May 1977: 175-180.

Trend Reports and Technical Digests

11

1. Trend Reports

1.1 Introduction

Trend reports are regarded as useful information consolidation products especially for the subject specialists and decision makers. They provide systematic view of recent developments and current research trends in a subject field during a specific period of time, usually ranging from two years to five years or more. The trend reports are usually prepared by information specialists with assistance from subject specialists and written with help from technical writers to ensure good language.

1.2 Definition

According to Gopinath "A Trend report is an exposition of a subject, giving an account of the general direction of research in the subject, based on a review of the documents on current developments. The trend report service is designed to help the specialist reader in the productive utilization of his time and in the conservation of research potential. The training of a documentalist equips him to do this service satisfactorily. However, the specialists' cooperation is necessary."

1.3 Purpose

The main purpose of trend reports is to briefly capture the major trends in a given area of research based on the analysis of literature on the given subject published during a specific period of time.

1.4 Need

The present day scientists and other researchers are facing problems of rapid growth of information due to increasing number of available documents. The current awareness services being provided by the information specialists are not adequately meeting their rising needs. The trend reports, however, provide an overview of the current developments and research trends in a particular subject. Trend reports fulfill the following needs:

(a) To help researchers keep themselves abreast of latest developments in a particular subject field, identify new areas of research and utilize their time and effort more productively towards research;

(b) To help planners and decision makers and funding agencies to sponsor a research; and

(c) To help students to identify an area of research.

1.5 Preparation of Trend Reports

Preparation of trend reports is not an easy task; it is rather a time consuming and complex process. Earlier, the trend reports were prepared by subject specialists based on the documents available in their institution's library/information centre. It involved great deal of time and effort in reading, sifting, evaluating and selecting the documents to be included and then writing the report. In terms of intellectual effort involved in writing a trend report it was no less challenging than the original research paper/s.

Meanwhile, new techniques of data collection and availability of electronic sources of information in the libraries and information centres have made it possible for library and information professionals to take up the task of writing trend reports.

Library and information specialists have the necessary skills, knowledge and competence to analyse a subject in great depth and represent it in a form to bring out the appropriate strength and relation between different components of a subject in a helpful way. Equipped with the professional knowledge, they can view the subject of trend report in proper perspective of the development of a subject in past, present and near future. They can easily make use of the guiding principles derived from the library and information experience, in the presentation of ideas in a helpful sequence. In recent times, some of the information professionals working in environments of special libraries and information centres are also getting training in technical writing, thereby preparing themselves for editing, formatting, and giving physical shape to the trend reports.

Library and information specialists should posses the following skills and qualifications for writing of a trend report:

(a) Familiarity with different components of the subject on which trend report is to be prepared.

(b) Proper rapport with the specialist users to find out subject approach and views on the subject of the trend report.

(c) Knowledge of the exact subject requirements of specialist users.

(d) Good knowledge of different sources of information on the concerned subject from which material is to be collected.

(e) Familiarity with technical writing skills and methods of compiling a trend report.

The entire work involved in the preparation of a trend report should be divided, for convenience, into three planes of work as suggested by Ranganathan, viz.,

- the Idea Plane
- the Verbal Plane
- the Notational Plane

The work involved in three Planes of work has already been discussed in an earlier chapter.

1.6 Methodology for Preparation of a Trend Report

The methodology and steps for preparation of a trend report are similar to that of the preparation of a state-of-the-art report. However, these can be briefly outlined below:

1.6.1 Selection of Specific Subject

To begin with, the choice of the specific subject has to be made on which the trend report is to be prepared. Library and information professionals have to deal with many queries/ issues about the subject so that they understand well the highways and byways of the subject.

1.6.2 Survey of Users

They must be familiar with the specific requirements of the target group of users. This may be based on the user surveys and the analysis of their information needs.

1.6.3 Collection of Data

It is a necessary prerequisite for preparing a trend report or for that matter for any ICP. It implies the selection and acquisition of various information sources and cull out requisite data or information from the documentary sources including databases, electronic sources, CD-ROMs, Internet, etc.

1.6.4 Appraisal of Information

The library and information specialists, as authors of the trend report, would scan through the collected information for its relevance or otherwise. They would have appraisal of information regarding its quality, scope, appropriateness, time lag, etc.

1.6.5 Arrangement of Information

The arrangement of information and other data should be based on its helpfulness to the users for whom the trend report is being prepared. This depends upon the purpose and specialization of the users.

1.6.6 Integration of Information

The next step is to integrate the arranged data or information into a well organized text. For this purpose, the intellectual level of the users, choice for a particular mode of presentation of information, etc. need to be considered.

1.6.7 Packaging

The integrated or restructured information is now packaged in a form that is acceptable to the users. The use of terminology, technical or common language, has to be according to the standard of the users. The first draft has to be checked thoroughly not only by information specialists but also the subject specialists.

1.6.8 Repackaging

It deals with the presentation of information in more understandable, readable, acceptable form. This is like preparation of the second draft of the trend report at which stage care must be taken regarding the language, technical writing of the report, etc.

1.6.9 Finalization of Report

At this stage, the writer of the report on the basis of the feedback regarding addition or omission of information, or changes in the sequence of information presented in it. Besides, the report also needs to be checked for headings, subheadings, placement of tables, illustrations, references, etc.

The final product contains the following parts:

(a) Title page
(b) Contents page
(c) Short informative abstract
(d) Body of the text, chapter wise
(e) Expressive index
(f) Bibliography or references
(g) Appendix giving the definitions of the technical terminology.

1.6.10 Printing and Distribution

Once the trend report is finalized, it is sent for printing a particular number of copies which are thereafter distributed among the specific segment of the users.

Examples

1. *IFLA Trend Report 2012.*
2. Eisenberg, Michael B., et al. *Trends and Issues in Library and Information Science.* ERIC Report, 1990.
3. Tiwari, S.K. *Recent Trends in Wind Energy.* New Delhi: INSDOC, 1982.
4. World Bank. *Trends in Developing Economies.* Washington: World Bank, 1995.

2. Technical Digests

2.1 Introduction

For industrial growth of a country, scientific and technical information, marketing and commercial information are its life blood. But the very rapid growth and proliferation of all information related to it makes it very difficult for the people in industry to keep themselves abreast of the latest developments about the new products, machinery, manufacturing processes, management techniques, etc. It is, therefore, essential to meet their information requirements through a specialized service called technical digest service. It can bring out an ICP, Technical Digest, which analyses, consolidates, evaluates, and disseminates the latest technical information to specialists in industry.

2.2 Definition

A digest is a systematic condensation of several written works, often prepared by someone other than the authors of the original works. It is larger than a synopsis, and is prepared with headings, subheadings to facilitate quick reference.

Webster's Third International Dictionary defines a digest as "to compress a piece of literature or a body of information into a short summary form, containing essential core of the matter."

According to Guha, "digest is actually a fuller representation of a document, rewritten for a purpose or to suit the requirement of different group of people, but intended to serve as a complete substitute for the original document."

Sometimes, the term digest is synonymously used with compendium. It can, therefore, be defined as "Body of information or written matter involving information consolidation by condensation. It is a compendium of information gathered from various sources, systematically and methodologically arranged, classified under headings and subheadings. It is prepared either on demand or in anticipation for quick and ready reference, with subject scope spanning from literary to science and technology."

A technical digest is also not as different from a digest as defined above, except the subject scope. A technical digest includes in its scope mainly science and technology, and management aspects of a subject. According to Guha, "Technical digest is a product of condensation process. It performs as accretion function. It may be directed or oriented to various groups of users, viz., managers, technicians, operators, etc."

2.3 Need

In industrial environment, technical digest is a useful source of information for managers, technical workers, etc. as they can obtain information about technical know-how and technical data helpful in their work.

In industry, different categories of personnel and workers have different types of information requirement. For example, managerial personnel require product-oriented information such as technical, commercial and marketing information. On the other hand, technical operators and workers require information on solving technical problems, new ideas and processes, which can help them in day to day working. Therefore, the contents of technical digest should also be structured according to their levels and needs so that managers can save their time in decision-making, and other workers get their requisite information and put in more effective and efficient work.

2.4 Functions

Technical digests perform the following functions:

(i) To keep various levels of personnel in the industry abreast of the latest developments in their respective fields of interest.

(ii) To provide timely and reliable information to keep users aware of the new developments taking place in their areas of specialization.

(iii) To disseminate latest technical know-how.

(iv) To help solve technical problems by providing new ideas and information about new processes in day to day working.

(v) To serve as an efficient channel between research and production centres.

2.5 Categories

Keeping in view the various levels and types of personnel working in industry, the technical digests are prepared for each of these categories. For instance:

2.5.1 Top Management

The digests for top management usually cover information on the following topics:

- Corporate Management and Planning
- Finance
- Production
- Research and Development
- Personnel
- Sales
- Public Relations
- External Regulations

2.5.2 Middle Level Management

The digests meant for the middle level or supervisory management should contain information about:

- Enhancing utilization of installed capacity and resources,

- New production processes and/or techniques helpful in maximizing production,
- Problem-solving and decision-making.

2.5.3 Operator Level Professionals

For operators and other workers, the digests should contain information which they can easily understand and put to use. Such digests help them in:

- Solving day to day technical problems,
- Ensuring modernization of existing production methods and techniques,
- Helping in small improvements in manufacturing/ process techniques leading to major savings,
- Assisting in mechanization of existing work.

2.6 Methodology for Preparation of Technical Digests

For systematic planning of technical digests the methodology to be followed is by and large, the same as for other information consolidation products. However, some of the important steps are discussed below:

2.6.1 Identification of Users and their Information Requirements

The first step is to know about the users as they are the most important element in the process of preparing a technical digest. The subject interests, professional knowledge, activities, functions, levels and responsibilities of the users in the organization need to be known. As potential users in industry work at different levels—like managerial, supervisory, operational, etc. and in different divisions—like Production, R&D, Sales and Marketing, etc. they would require various types of information depending upon the nature of their work. Technical digests are prepared to meet the specialized requirements of different personnel/users.

2.6.2 Sources of Information and their Collection

It is important to identify and collect various sources of information for preparing technical digests. These information sources include science and technology journals, trade journals, manufacturers' catalogues, newsletters, organizational publications, company reports, company websites, databases, trade portals on the Internet, etc. Besides, it also covers the latest sources on the new industrial products, processes, machinery and materials.

2.6.3 Information Analysis and Consolidation

Once the requisite information from various sources as mentioned above is collected it need to be analyzed, evaluated and consolidated for compiling a technical digest. The information analysis and consolidation has to be made keeping in view the requirements of user groups working at different levels, i.e. managerial, technician, etc. Analysis and consolidation of information is basically intellectual activity, hence it needs to be carried out carefully by subject specialists and information specialists.

2.6.4 Evaluation

Evaluation of consolidated or restructured information is an important step before it is communicated to the various user groups. The information specialist entrusted with the responsibility of compiling the digest is supposedly has sufficient subject knowledge so as to thoroughly evaluate information to be incorporated in the digest. He has to check the validity and authenticity and reliability of the sources of information.

2.6.5 Presentation

Presentation of information in technical digest is an important stage of its compilation. Keeping in mind the target user groups, the language to be used in technical digest should be simple and familiar. The contents of this product should be accurate, authentic, and brief. Other qualities of good technical digest include clarity of presentation, consistency, precision of information, etc.

2.6.6 Layout

Besides presentation, the layout of a technical digest is equally important. The following points may be kept in mind for the layout:

– Should have appropriate eye-catching headings,

- Should be short within 200-400 words and must not exceed 3-4 typed pages,
- Should include sources of information with complete bibliographical details at the end of the digest.

2.6.7 Dissemination

At this stage the technical digest is ready for distribution or dissemination to the target audience without any further delay. To make the digest useful time lag between the collection of information, its preparation, and its dissemination should be minimum. This is necessary so that the data and facts do not loose their value to the users.

Examples

1. *Management Digest*. Bimonthly. Hyderabad: Small Enterprises National Documentation Centre.
2. *TERI Information Digest on Energy and Environment*. Quarterly. New Delhi: Tata Energy Research Institute.
3. *Science Education Digests*. ERIC, 2000.

References

Atherton, P. *Handbook of Information Systems and Services*. Paris: UNESCO, 1977.

Guha, B. *Documentation and Information: Services, Techniques and Systems*. 2nd rev. ed. Calcutta: World Press, 1983. Reprinted 1999.

Neelameghan, A. *Presentation of Ideas in Technical Writing*. New Delhi: Vikas, 1975.

Saracevic, T. "Processes and Problems in Information Consolidation". *Information Processing and Management*, Vol. 22 (1) 1986: 45-60.

Saracevic, T. and Wood, J.B. *Consolidation of Information: A Handbook on Evaluation, Restructuring and Repackaging of Scientific and Technical Information*. Paris: UNESCO, 1981.

Seetharama, S. *Information Consolidation and Repackaging: Framework, Methodology, Planning*. New Delhi: Ess Ess Publications, 1997.

Subramanyam, K. "Review Literature." In *Encyclopedia of Library and Information Science*. New York: Marcel Dekker, 1979. Vol. 26, pp. 461-74.

Information Consolidation Units: Planning and Management

12

1. Introduction

There has been exponential growth of literature and information pollution in recent years, especially in the subjects of science and technology. Information has not only appeared in various forms but in various kinds of documents also. Such a development has made it well nigh impossible for specialist users to analyse and evaluate all published but relevant information critically. These and other developments have constrained the library and information professionals to pay more attention to the individual needs of users. The information specialists have, therefore, designed and developed many information services and products for the different segments of user groups. The specialist libraries and information centres have set up Information Consolidation Units (ICUs) for providing the evaluative and critical information to meet the various types of needs relating to their institutional activities. In other words, the ICUs were to support the research and academic activities of the parent organizations as well as their scientific and technological, managerial and general objectives. In order to achieve these objectives, analyzed, evaluated and consolidated information on different aspects of various subjects has to be made available. Therefore, a need arises for the establishment of a well-staffed and organized Information Consolidated Unit.

2. Management of ICU

There are various ways of defining management, and one of them, and perhaps simpler one, is "Getting results through people". In other words, management guides human and physical resources into dynamic organization units to attain their objectives of the satisfaction of those served.

Management is the process by which decisions are made and carried out within an organization. Management can be broadly defined as working with people to determine, interpret and achieve organizational objectives by performing the functions of planning, organizing, staffing, leading and controlling.

In a more specific way management can be defined, according to Saracevic, "as the guiding of human, physical and economic resources into dynamic organization units that attain their objectives to the satisfaction of those served, and with a high degree of morale and sense of attainment on the part of those rendering the service."

Thus, management is a collection of processes such as decision-making, problem-solving and action-planning. These processes involve the management of resources, viz., human, financial, material and time.

As most of the ICUs are located within the parent organization, the objectives and framework of parent organization define the objectives, role, financing, staffing, and operations of an ICU. Thus, all management functions of ICU are dependent on the management of the parent organization.

Management of ICUs would cover the entire spectrum of management activity. But for our purpose, according to Seetharama, it would include such components as policy formulation, planning, control, resource allocation, marketing of products, etc. A discussion on these components may help to understand them.

2.1 Policy Formulation

While framing policies, as guides to managerial decision-making, care should be taken to base them on the requirements and activities of the ICU. It is aimed at improving its operations, programmes of services, and resources on a long-term basis.

According to Atherton, policies are restrictive and permissive, spelling out limits of acceptable action, and granting freedom to operate independently within these. The policy once framed enables the library and information professional for long-range management and planning.

2.1.1. Process of Policy Formulation

Atherton has suggested a framework for policy analysis and development that includes several tasks as given below:

(i) **Initiation of Policy Development Process:** It involves plans of action.

(ii) **Review of Existing Policies:** It involves identification of existing formal and informal policies, if any.

(iii) **Analysis of Existing Policies:** It involves evaluation of existing policies for their effectiveness.

(iv) **Recommendations Regarding New and Revised Policies:** It involves formulation of new and revised policy statement for the major areas of concern not covered or inadequately covered.

(v) **Process of Formulating Policies:** A schedule for formulation of new policies may be prepared indicating priority list of areas to be covered.

(vi) **Preparing a Policy Manual:** It will include the guiding principles behind the operating design, such as format of policy statement, topical presentation of policies, etc.

(vii) **Dissemination of Policy:** Soon after establishing policies information about them should be disseminated.

(viii) **Periodic Policy Review:** It involves review and updating of policy to meet the changing needs and objectives of the organization.

2.1.2 Areas of Policy Development

Given below are the areas on which policy statements should be developed.

(i) Collection of Material: It may include policy matters about types of materials, relation of collection to research, responsibility for decisions on selection, involvement of users in collection development, size and proportion of resources allocated, etc.

(ii) Services and Access to Materials: Policy matters may cover users to be served, provision of physical facilities, user orientation, scope and extent of services, decentralization of services and access to materials, etc.

(iii) Staff Resources: It may include organization of human resources, categories of capabilities, provision of specialized skills, status and role of manager, development of staff skills, attitudes, professionalism, etc.

(iv) Overall Management Considerations: It will have policies about administrative and working relationships within the parent organization, utilization of limited resources, cooperative relationships with other information services.

It becomes obvious from the above that policy decisions in these areas can guide both the information consolidation service in fulfilling its functions and the organization in fulfilling its expectations.

2.2 Planning

It is an acknowledged fact that fast dissemination of information resulting from scientific and technological research leads to faster pace of national development. Such awareness on the part of national planners, scientists, technologists, decision makers, etc. have led to drawing up plans for scientific information centres for various subjects at the national and regional levels. However, for the planning and development of ICUs desired attention has not been paid by the decision makers.

In order to meet with the objectives of service to users, the ICUs must be well planned in the absence of which they will remain poorly organized and ill-equipped.

2.2.1 Benefits of Planning

(i) Planning gives direction to growth and complexity;

(ii) It minimizes ad hoc decisions; and

(iii) It provides a basic framework for service.

2.2.2 Systems Approach to Planning

Seetharama has suggested that while developing a systems approach to planning of ICUs, the following parameters may be considered:

(i) The overall objectives of the ICU;

(ii) The components and their characteristics;

(iii) Functional organization of the components into a coherent whole;

(iv) The interlinks and relations among the components;

(v) The characteristics of the environment in which the ICU is to operate from time to time;

(vi) The built-in control, preventive and corrective measures which would enable the system to adapt to the changing information environment and management policies and directions;

(vii) The conditions for healthy growth and development of the ICU; and

(viii) Identifying the agency which would ensure continuous inputs, management, evaluation and adaptation of the systems along right lines.

2.2.2.1 Model for Planning

A model for planning an ICU may be developed which would include three major elements as explained in Fig. 12.1.

1. Strategic Planning: This stage of planning deals with future opportunities, problems, and alternative courses of action.

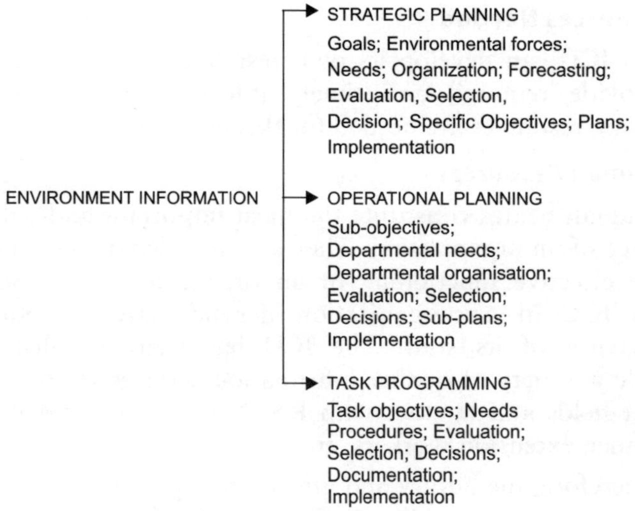

Fig. 12.1: Total Planning Framework

2. Operational Planning: At this level, it is the linking of organizational components (i.e. techniques, materials, procedures, strategic plans, rules, people, capital and facilities) to facilitate the flow of interrelated events. The emphasis here is on getting things done through people.

3. Task Programming: This stage has plans of specific tasks (i.e. a particular event, or a specified activity performed in relation to the tools, materials, equipment and technology used) to be carried out where the emphasis is on technical activities.

2.2.3 Basic Elements for Planning

Given below are the basic elements of the ICU which should be carefully planned and developed.

(i) Information sources or database;

(ii) Information consolidation services;

(iii) Manpower for the operation and management of the ICU;

(iv) Furniture and equipment;

(v) Building; and

(vi) Finance.

3. Resources Needed

An ICU can develop its own resource collection in order to provide required and timely information services. The following resources are needed for this purpose:

3.1 Human Resources

Human beings constitute the most important and valuable resource of an organization. They are regarded as the vital key to the effective functioning of an organization. The success of an ICU in an organization depends largely upon the competence of its staff. The ICU has been established to provide a variety of quality information services in specialized subject fields and interact with R&D personnel, management personnel, extension workers, etc.

Therefore, the library and information professionals should have high professional (i.e. in library and information science) and subject (i.e. in science, engineering, etc.) qualifications, adequate experience and training, innovative capability, proper attitude toward work as well as users.

In an ICU, to man it well, at least two categories of personnel are required, viz. professional personnel, and supporting staff. The number of personnel at each level, their qualifications, etc. will depend upon the type, quality and quantity of work to be done in each ICU. But it must be according to the national and international standards for estimating and recruitment of manpower.

According to Seetharama, manpower resources for an ICU should be examined from the point of view of:

- Determination of staff
- Skill requirements and levels of training of staff
- Availability of staff at different levels
- Selection of staff
- Cost estimates and salary structure

3.2 Manpower Requirements

Many ICUs are small; some are even operated by only one person. The staffing requirements mentioned below may have to be concentrated in one or few persons. Thus, these requirements may have to be translated into required expertise rather than treated as a required number of persons. The manpower requirements are:

(i) a director having management capabilities, and a visionary outlook;
(ii) professional(s) with information expertise;
(iii) professional(s) with subject expertise (internal or external consultants); and
(iv) supporting technical personnel as needed for given technical jobs.

3.2.1 Manpower Training and Development

Education, training, acquisition of skills, etc. are learning processes that help an organization to improve the abilities and competencies of its manpower so that they perform better and deliver high quality products and services. Therefore, manpower training is a necessity.

Training is a process which personnel in an organization are offered to enable them to absorb some new perspectives, understanding, value, attitude, technique, or skill with reference to activities and programmes of an organization. Training activities and programmes flow from the analysis of need for it in the parent organization. In the field of library and information science, change of perception, attitudes, approaches, etc. has become extremely necessary, particularly in view of increasing application of information technology.

Manpower training, therefore, requires a continuous programme of professional development. This can take the following forms:

- In-house training;
- Sending professionals to outside courses, conferences/ seminars, etc.;
- Cooperation with professional associations;
- Subscription to professional literature.

3.3 Technical and Technological Resources

In an organizational set up the effectiveness of decision makers depends to a considerable extent upon the type of information made available to them. The use and application of information may yield desired results if processed and disseminated for use with the help of tools of modern technology. This technology, also called as Information Technology in this context, according to Neelameghan, facilitates collection, storage, organization, processing, analysis, presentation, communication, and dissemination of data and information for decision-making using mechanical and/or electronic means such as computers and telecommunication equipments is indeed an essential support to development.

Saracevic and Wood also made a similar case for Information Technology, and pleaded that among other things, organizing mechanisms for more effective information delivery to problem solvers and decision-makers employing information technologies and networks appropriate to their circumstances.

Rowley reiterated that "Information technology means the collection, storage, processing, dissemination and use of information. It is not confined to hardware and software but acknowledges the importance of man and goals he sets for this technology, the values employed in making these choices, the assessment criteria used to decide whether he is controlling and being enriched by it."

Information Technology therefore, is electronic technologies used for collecting, storing, processing and communicating information in all its forms.

ICUs can play significant role in the development of information technology by providing appropriate information to the decision-makers. It is, therefore, necessary for the ICU to explore the technical resources available at local, national and international levels. In this context, it can develop its capacity by computer and communication capabilities so as to achieve the maximum productivity in its operation.

3.4 Physical Resources

An ICU requires the following items in terms of physical resources:

(a) Furniture articles

(b) Equipments

(c) Space facilities/or separate building, etc.

3.4.1 Furniture Items

The various items of furniture and equipments required for an ICU depends also on

- The total document collection,
- Number of users, and
- Number of professional and non-professional staff.

Regarding the selection and purchase of major items of furniture and equipments, the Indian Standards developed for this purpose may be considered. These include the following:

1. IS: 1829 (Part 1)—1961. Specification for library furniture and fittings. Part 1: Timber. (First revision, 1978).
2. IS: 1829 (Part 2)—1993. Library furniture and fittings, Part 2: Steel-specification (first revision).

3.4.2 Equipments

Among the equipments needed for an ICU are included computer/s and data processing, fax machine, duplicating/ copying machine, steel cupboards, filing cabinets, microfilming machine, audio-visual aids, miscellaneous office items, air conditioning equipments, etc.

3.4.3 Space Requirements

In order to carry out its functions efficiently, ICU needs sufficient space either in the same building or at a new place. For calculating the space requirements, the following Indian Standards may be helpful:

1. IS: 1553-1960. Code of practice relating to primary elements in the design of library buildings. (Second revision, 1992).

It provides information about the average size of different libraries regarding number of books, bound periodicals, current periodicals, number of seats for readers, and strength of staff.

In case an ICU has to be located separately, remarks of Kertesz are worth noting: "Experience gained indicates that a small centre does not operate in a complete isolation. For example, in an office building without neighbours engaged in a related technical activity, it has been recommended to establish a new centre at a location 'where relevant science is flourishing.' Potential homes include major libraries, document collections, government agencies and editorial offices." He further stated that "The most promising site would be a large, multidisciplinary research institution, a university or industrial laboratory, where interaction with colleagues is conveniently available."

3.5 Information Resources

About two decades ago, information sources were synonymous with the printed works. Today the definition is turned on its electronic head, observes William Katz. Basic reference sources are available online or as CD-ROMs. It is believed that soon many sources will not be available in print anymore.

The various information services and products to be designed and developed by an ICU usually depend, inter alia, on the strength of its sources of information in print and/or electronic version. However, the two well-known categories of information sources are:

1. Documentary Sources

Documentary sources of information are further grouped into three major types:

1.1 Primary Sources: These are the first formal published records of original research and development and constitute the latest available information which is neither interpreted nor condensed nor evaluated by others. These include such sources of information as research journals, research monographs, research reports, conference and seminar proceedings, patents, standards, theses and dissertations, autobiographies, trade literature, unpublished sources like diaries, letters, etc.

1.2 **Secondary Sources:** These are either compiled from or refer to primary sources of information. They describe, interpret, analyze and evaluate the primary sources and at least one step removed from the information they refer to. This category includes bibliographies, yearbooks, handbooks, dictionaries, encyclopedias, indexing and abstracting periodicals, biographical sources, geographical sources, literature reviews, treatises, etc.

1.3 **Tertiary Sources:** This category of sources includes information distilled and collected from primary and secondary sources. These include materials in which information from secondary sources has been digested, reformatted and condensed to put it into a convenient, easy-to-read form. For example, fact books, bibliographies of bibliographies, directories, guides to literature (i.e. bibliographies of reference sources), etc.

1.4 **Electronic Sources:** The use of information technology has also resulted in the production of electronic sources of information. Several language dictionaries and encyclopedias are now available in CD-ROM format which can combine text, images and sound in one package. The online databases, online catalogues are also available which can be acquired in the ICU.

2. Non-Documentary Sources

Non-documentary sources of information, though different from documentary in nature, yet are regarded as quite useful and helpful. These are broadly of the following two types:

2.1. **Human Resources:** These are unique and invaluable sources of information, in human experts, carrying expert knowledge and/or skills. These persons may be working in different institutions but are associated with information or knowledge in different subject areas. They may be information generators, information gatherers, information processors, information recorders, information disseminators, information retrievers, and information technologists. An ICU must become acquainted with various types of human experts, their work and activities, etc.

2.2. **Institutional Information Sources:** Many human activities like education, research, health care, business and industry, defence and security, cultural activities, etc. are pursued through various types of institutions. Every institution builds up its own resources and develops its own products and services by generating information/knowledge of various kinds. Their records in the form of journals, reports, special publications, etc. are communicated to the public but much information still survives in the files which remains unnoticed but is useful. An ICU, therefore, must become acquainted with various types of institutions, their nature of work and activities, and the material available from them.

3.6 Financial Resources

Financing is the most critical and vulnerable aspect of Information Consolidation Units (ICUs) established either within organizations or independently, as they are the spending units for the organizations. Information consolidation is a costly proposition which requires adequate and sustained financing by the parent organizations.

ICUs being information service organizations, cannot easily demonstrate tangible benefits being derived by their users. Therefore, funds allocation to the ICUs does not come easily. These units are rather asked to reduce expenditure to the hilt.

Their situation is well summed up by W.M. Carlson, an American defence official, when he remarked: "Information is a resource whose generation consumes time and money, whose use conserves time and money, whose cost and value are not known to management." It makes it amply clear how difficult it is for the ICUs to make demand for funds from their respective managements.

The financial requirements for the establishment, operation, and maintenance of an ICU should be worked out in a feasibility study. It is necessary to analyse costs and to determine sources of financing. According to Pauline Atherton the cost analysis may be based on assessment of expenditure by purpose and can be classified into two categories, capital and current:

(a) **Capital Expenditure:** Expenditure for the acquisition of site, building, equipment, furniture, permanent collection of materials, vehicles, etc.

(b) **Current Expenditure:** Salaries of personnel, running costs (electricity, communications, etc.), maintenance costs (building, equipment, furniture, transport vehicles, etc.), expenses for materials and supplies, miscellaneous expenses (insurance, hiring of buildings, etc.).

The financial requirements for the basic elements, viz., information sources, information consolidation services/ products, manpower, furniture, equipments, building, etc. should be calculated individually so that the full financial plan could be prepared.

For different activities of an ICU, one can work out financial requirements at different stages. Saracevic and Wood have mentioned the following requirements:

(i) Planning Costs
- Feasibility studies
- User studies-marketing research and analysis
- Design

(ii) Launching Costs
- Development of functions, products, and services
- Deployment and testing
- Development of market and marketing
- Investment in facilities, tools and equipment
- Staffing and staff training and education

(iii) Operating Costs

- Salaries for a Director, professionals and supporting staff
- Fees for consulting experts
- Cost of procurement or access to necessary information sources
- Expenses incurred in given functions
- Marketing expenses
- Continuing education expenses for staff

To cover all these expenses the financial outlay can be considerable. Therefore, it is not unusual to seek and establish composite funding of the ICU.

Saracevic and Wood have suggested that for the planning and launching costs, financial assistance from national and international organizations (outside of direct sponsoring or host institution) is indispensable. Such assistance is necessary for a period of one to three years. A good proposal is required to secure the assistance, which in turn requires considerable and careful preliminary work.

Further, it is suggested that for the operating costs, the ICU has to rely on financial assistance from the sponsoring or host institution. This support may be either partial or complete. But with increasing emphasis on self-sufficiency in recent times, ICU may have to resort to cost recovery programmes like getting revenue from sales of services/products. This recovery of costs may be partial or complete. When cost recovery programmes are instituted, careful consideration has to be given to the following:

- Pricing policies and price elasticity;
- Spreading the operating costs over a sufficiently large number of users, so that economy of scale may be working;
- Strictly controlling and minimizing internal expenditure and maximizing the efficiency of operations;

- Pooling of resources with other information systems; establishing a resource—and operation-sharing.

But Seetharama asserts that such cost-recovery programmes may be workable in industrialized countries, it would not be so in developing countries. This is because of the fact that the users, in developing countries, cannot afford to pay cost-recovery prices and even the few who may afford will have problems in securing foreign exchange, he says. This has been the experience of the Asian Institute of Technology, Bangkok, after nearly first ten years of its experience with its several specialized information analysis centres. In his Report M. Brandreth found that they will probably never recover more than half of the annual operating costs.

3.6.1 Suggestions for Seeking Finances

In view of the above-mentioned situation, some suggestions have been made by Seetharama for seeking finances:

- From donor agencies like IDRC, Unesco, etc. who support information consolidation activities;
- By creation of *endowment fund* whose interest would cover some or all of the operating costs;
- By establishing *multi-donor fund*;
- By tapping the private sector;
- By including an information component in the budget of every research project;
- By straight forward charging of fees for services rendered;
- By membership subscription;
- Charges for documents—such as, periodic bulletins, specialized journals, state-of-the-art reports, compilations, data sheets, bibliographies, etc.;
- Charges for translation;
- Charges for reproduction of documents;
- Charges for responses to enquiries;
- Charges for symposia, seminars, and training sessions.

It would be appropriate to mention here that viable functioning of ICU cannot depend solely on direct subsidies, or on a system of user charges, or on efficient operations and cost-sharing arrangements. There must be a careful balance of all these factors, depending on the individual situation.

Pauline Atherton states in this connection that the most important and obvious criterion is how well the ICU meets its goals and how it performs the services it was set up to provide.

4. Organizational Structure of ICU

While developing the organizational structure of an ICU, the information managers should keep in mind its functions, activities, and programmes. The component units of the ICU should therefore work in cooperation and coordination of each other to give it holistic picture. The organizational structure must be clear, complete and integrated so that it would facilitate effective control of the various activities of the ICU. Seetharama suggested that it may be worthwhile to consider the ICU as composed of a Technical Library Unit, Information Service Unit, Information Use Promotion, and Common Facilities Unit.

The broad functions of the different units of the ICU are indicated in Fig. 12.2.

The functional sub-units of the ICU as mentioned in the figure are only for the convenience of management, and they should not function as watertight compartments. Their sole objective should be the achievement of maximum productivity in service and management. This requires careful planning and coordination of the work of the different sub-units of ICU. The feedback from reference and information services, liaison, for example, can provide each sub-unit with an idea about the information needs and services preferred by different user groups. This will help each sub-unit to develop tailor made work and services to effectively meet their information needs. Similarly, the scanning of documents and extraction of information is preliminary work common to the provision of current awareness services, preparation of abstracts, technical notes, digests, etc. It is important that the work done for each of the services should be coordinated in such a way that the repetition of the same or similar steps involved should be reduced to the minimum effort. For instance, in the production of different services the scanning of documents and extraction of information should be done only once and not be repeated for every service.

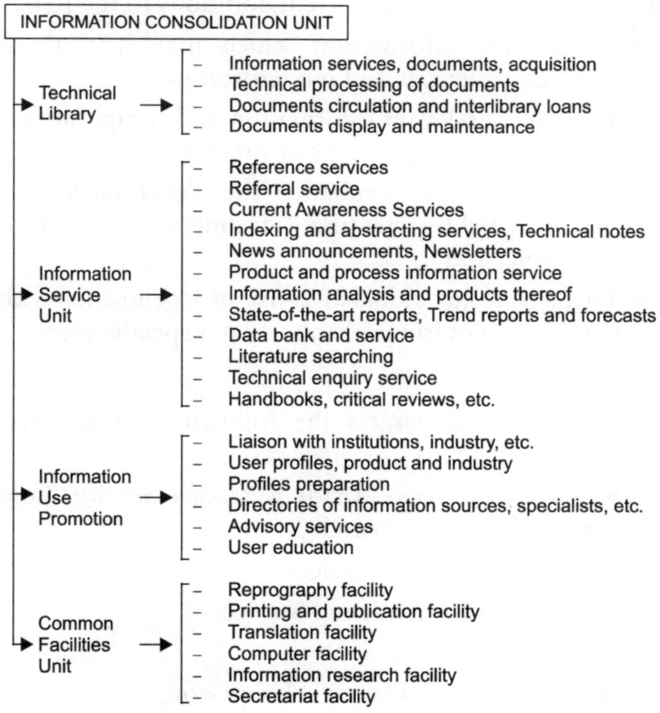

Fig. 12.2: Organizational Structure of ICU

5. Dissemination and Communication

The job of an ICU has not been completed with the institution of a service or developing of a product. Active dissemination has to be included as an integral part of the Unit or the system as a whole. Dissemination or communication of information is a process of conveying information through various channels—oral or written, formal or informal—to users.

5.1 Purpose of Dissemination

The purpose of dissemination or communication of information may be one or more of the following:

(a) To keep the users abreast of current developments in their areas of interest;

(b) To inform about the current additions to the ICU;

(c) To provide information which may help to solve technical advances and modernization;

(d) To provide relevant information and comparative data about particular products or processes;

(e) To review (and evaluate) the developments in a specific field—scientific, technical and techno-economic; and

(f) To provide an exhaustive list of documents with or without annotations/abstracts on a specific topic.

5.2 Services of ICU

An ICU usually generates the following services for its users:

(a) Reference Services—Ready Reference Service, Referral Service, Literature Search

(b) Current Awareness Services

(c) Indexing and Abstracting Services

(d) Compilation Services

(e) Special Services to Specific User Groups

(f) Supporting Services—Translation Service, Reprographic Service, etc.

5.3 Products of ICU

The following information consolidation products can be listed that an ICU can produce for its users, though some of them may not be produced in developing countries.

(a) New announcements, newsletters

(b) Pamphlets, brochures, question-answer sheets

(c) Monographs, technical reports

(d) Critical reviews, state-of-the-art reports, handbooks

(e) Technical guides and "how-to" booklets

(f) Operational or maintenance manuals; specifications; standards

(g) Recurring summaries of advances in a topic; short textbooks

(h) Market reports; industry reports

(i) Statistical condensates

(j) Briefings

(k) Radio and TV shows

(l) Comic strips, photo novels

(m) Wall posters

(n) Lectures, tutorials

(o) Demonstrations, exhibitions

(p) Video newsletters; audio-video presentations

(q) Videotapes, videodiscs for training

(r) Computer online searching

(s) View texts, tele texts, etc.

5.4 Dissemination Channels

A dissemination channel is defined as the means of transition or mechanisms by which information consolidation products are distributed and delivered to users. Saracevic and Wood have listed some of the important dissemination channels as follows:

(a) Interpersonal delivery—products personally delivered to each user.

(b) Group personal delivery—products given to users at meetings, demonstrations, etc.

(c) Strategic placement—products put at key sites for users to pick up, etc.

(d) In-house dissemination.

(e) Local depository—products circulated from a library, extension service, etc.

(f) Mass media—print.
(g) Broadcasting—radio, TV.
(h) Mail and other means of public distribution.
(i) Telecommunications—telephone, satellite, etc.
(j) Computer networks—online systems, computer conferencing, etc.

The above-mentioned distribution and communication channels have their own advantages; the users may have their own preferences, depending on the nature of their work and field of specialization. However, some factors such as cost, effectiveness, and timeliness must be considered while selecting different delivery systems.

5.5 Life Cycle of ICU

Like human organizations, an Information Consolidation Unit too has different stages in the cycle of its life: from inception, to birth to adolescence, to adulthood and possibly to death. J. Valls has made out a case for different requirements for each phase.

(i) Preplanning or Idea Phase (Inception): It covers the period from the time that someone has the idea for establishing the unit until a proposal for assistance is sent to a national or international organization. It requires:

– Knowing the existence of need and problem
– Knowing the subject/topic to be covered
– Types of users to be served
– Outline of the methods and approaches to be used
– Drafting of a pre-proposal
– Securing the backing of a host organization
– Searching for a funder for assistance

(ii) Planning Phase (Incubation): This phase covers the period from the time proposal is accepted to the time a unit is physically launched. It requires:

– Recruiting and training of staff

- Securing a location and facilities for easy access for experts and users
- Dealing with management aspects
- Conducting a user study and market analysis
- Establishing an Advisory Board and using it
- Getting the necessary tools and equipments
- Designing processes, products, services, and preparing specification for working procedures
- Planning of marketing, including evaluation of products and services
- Testing and adjustments
- Establishing connections with other information systems

(iii) **Launching Phase (Birth):** It includes the period extending from the announcement of the services until the time when the users start reacting. It requires:

- Effort to create a wide awareness of the Unit's existence and of the potential benefits to be derived from using its products and services.
- Promotion, user education, visits, demonstrations, etc. in order to achieve high visibility.
- Launching all operations and producing first products and services.
- Positive reaction to complaints, criticism.
- Continuing staff training.
- Reporting to funder.
- Showing results to sponsoring organization.

(iv) **First Operational Phase (Infancy):** It covers the period extending from the first user reaction to the provision of products and services and to the satisfactory execution of routine operations. It requires:

- Evaluation of user and market reactions
- Adjustments as necessary and reassignment of staff

- Increased attention to quality control in all operations, products and services
- Creating trust and confidence in users
- Strengthening connections with other information systems
- Expansion of marketing
- Consideration and development of new products
- Showing results to professional colleagues
- Showing benefits to sponsoring organization

(v) **Expansion or Second Operational Phase (Adolescence):** Covers the period extending from the attainment of full operations status to the achievement of a relatively high growth rate. It requires:

- Continuing evaluation, reevaluation and initiation of necessary changes
- Attention to continuing and intensified marketing and contacts with increased number of users
- Launching of new products and services
- Opening avenues to new users
- Seeking of new groups of experts on a rotational basis
- Rotation of Advisory Board

(vi) **Self-sufficiency or Third Operational Phase (Maturity):** Covers the period from achieving relatively full market potential (high growth rate) to the battle for continuing existence and survival, and to relative lower growth rate or even steady state. It requires:

- Continuing evaluation, reevaluation and adjustment to environmental conditions and changing user needs.
- Achieving financial cost-recovery programmes as planned, adjustments as necessary.
- Launching new products and services; improvement of old ones; mandatory abandonment of less useful products.

- Increasingly cultivating user contacts and trust; leaving nothing to routine with users.
- New training and professional development efforts for new staff; continuing education for all staff.
- Continuing marketing, but with new marketing approaches.

(vii) **Possible Discontinuance (Death):** An ICU can be terminated of its own accord because of inadequate budgets, performance, and/or internal failures. However, there are cases when the unit is closed because of other objective reasons such as:

- Sponsoring or host organization changed interests as to subject or topic and the information is no longer needed.
- The pace of the activity in a technical area of information coverage slackens, i.e., there is no information to consolidate.
- The information became available within reach of the user community on its own.
- The urgency of information needs has gone.

References

Atherton, Pauline. *Handbook for Information Systems and Services*. Paris: UNESCO, 1977.

Brandreth, M. *Specialized Information Analysis Centres in International Development*. Report of a Meeting held at Montebellow, Quebec, Canada, 4-8 October, 1982. IRDC Manuscript Report Series. March 1984.

Bryson, Jo. *Effective Library and Information Centre Management*. England: Gower, 1990.

Katz, William A. *Introduction to Reference Work*. 8th ed. New York: McGraw-Hill, 2002, 2 Vols.

Kertsz, F. "Collaboration Between IACs at a Large Multipurpose Laboratory". Quoted in Weisman, H.M. *Information Systems, Services and Centres*. New York: Becker and Hayes, 1972.

Rowley, Jennifer. *The Electronic Library*. 4th ed. London: Library Association Publishing, 1998.

Saracevic, T. *A Course in Information Consolidation: A Handbook for Education and Training in Analysis, Synthesis and Repackaging of Information* (Preliminary version). Paris: UNESCO, 1986.

Saracevic, T. and Wood, J.B. *Consolidation of Information: A Handbook on Evaluation, Restructuring and Repackaging of Scientific and Technical Information*. Pilot Edition. Paris: UNESCO, 1981. (PGI/81/WS/16).

——. *Guidelines for the Establishment of Information Consolidation Units*. Paris: UNESCO, 1984. (PGI-84/WS/19).

Seetharama, S. *Information Consolidation and Repackaging: Framework, Methodology, Planning*. New Delhi: Ess Ess Publications, 1997.

Valls, J. *Information Services for Developing Countries*. Bangkok: Asian Institute of Technology, 1983.

Webb, S.P. *Creating an Information Service*. London: ASLIB, 1985.

Knowledge and Skills for Information Analysis and Consolidation

13

1. Introduction

The human society of today is undergoing a sea change due to growth of information, the much quoted "information explosion", and certain other factors like information technology (IT) which is seen as "a cause as well as product" of this phenomenon. Therefore, it is considered, according to John Agada, as the "unrivalled change agent of modern times", which is transforming the information environment relentlessly. The information-intensive society has gradually emerged as what is termed as "Information Society" where problem-solving is made easier by utilizing existing pertinent information. This has resulted in new ways of performing old tasks quickly and more efficiently, new systems of information and information products, and a totally new work culture. There has been a tremendous growth in the use of computers and communication technologies which have considerably influenced handling, storage, processing, retrieval and dissemination of information to the users. But the ultimate aim is to make the right information accessible at the right time to the users to enable them to solve their information related problems. The information service professionals should take advantage of this new situation, and equip themselves in a way quite distinct from the conventional method, by acquiring new knowledge and skills.

2. Professional Knowledge and Skills

2.1 Need

Education is as important as any other resource for economic and industrial development these days because it is regarded as key to human resource development. Library and information centres have now moved from functioning as support centres to be come to be known as 'knowledge base' or 'knowledge centres'. To function as such they need trained and qualified manpower who could cope with the increase in the research activity of the parent and other institutions. This and the following changes speak volumes about the need for professional knowledge and skills for the library and information professionals.

1. As a result of the impact of IT, there has not only been instant access to information but the scholars, scientists, academicians, etc. have also been able to recognize the importance of information and the information professionals who make access to information possible.
2. There has been shift in forms of collection from books, periodicals, reports, conference proceedings, and other printed materials to non-print materials like floppies, CD-ROMs, videos, computer databases, etc.
3. A shift in library services has been another visible change. From reactive library services like circulation, reference, interlibrary loan, bibliographic services, etc. to proactive such as CAS, SDI, indexing, abstracting, database, etc. to active information based services that include specific databases, information analysis and consolidation, information reports, etc.
4. The users being more aware now have apparently changed their information seeking behaviour in which situation instead of quantity they now want quality information which should be timely. They want fast information access and result oriented retrieval of information.

5. As a result there has been a paradigm shift in the information professionals' approach to work. They are now transforming them from process oriented to user oriented, and reactive to proactive, from information gatherers to information analysts, and so on.

In this changed scenario, there is all the more need for acquiring requisite professional knowledge and skills which will provide them a broad understanding of the fields of knowledge of their organization. In the present information-intensive society for providing useful information services to various groups of users, the information professionals have not only to acquire appropriate knowledge and skills but also make it a life-long learning activity.

However, it may be pointed out that by acquiring the traditional knowledge about library and information science as a subject, which includes document selection, acquisition, classification, cataloguing, etc., certain information services like reference services, current awareness services, and literature search services can be provided to the different user groups working in different environments of organizations. These services are provided more efficiently and effectively with the use of information technologies, especially computer technology.

But Sweeney has suggested that apart from the traditional subjects taught in library schools, microcomputing, marketing, communication, technology and its applicability, information seeking behaviour, societal issues that develop from IT—copyright, privacy, database security, etc.—information management, economics of information, should be emphasized. In other words, as Seetharama observed, all knowledge required that would make the information professional think logically and clearly and that which would make him a fully integrated team participant and an efficient and effective information manager needs to be imparted.

Seetharama, however, argues that there are some information services which are not amenable as yet to information technologies. These services are known as information consolidation services and/or products. These include reviews, state-of-the-art reports, critical reviews, trend analyses, forecast reports, handbooks, monographs, technical manuals, etc. One of the reasons put forward for their not being amenable to IT is that the creation of such products/services involves the participation of several groups of specialists. He further observes that in this context the role of library and information science personnel becomes very important. And, it is for this reason that for playing this role successfully the information professionals need to acquire new knowledge and skills pertaining to information consolidation activities.

2.2 Selecting Appropriate Knowledge and Skills

Information professionals perform information work and provide information services to different user groups, at different levels, in different environments for which function they need to acquire necessary and sufficient knowledge and skills. Bhattacharyya has suggested a procedure for selecting required knowledge and skills for effectiveness of professional activities.

(a) Scan the predominant environment to ascertain the information needs of the users in different environments, and also the types of information services needed to meet these information needs;

(b) Mark out the professional work and services that call for advanced professional competence;

(c) Select the necessary and sufficient knowledge and skills that are warranted to display advanced professional competence;

(d) Mark out the areas that warrant further specialization at the advanced level considering the requirements of various environments;

(e) If it is warranted for the sake of economy, efficiency, and effectiveness, select the necessary and sufficient knowledge and skills essential for extending professional assistance to the professionals performing information work and services of advanced level; and

(f) If it is warranted for the sake of economy, efficiency, and effectiveness, decide about the necessary and sufficient knowledge and skills required for extending assistance to the professionals in their professionally routine tasks.

2.3 Appropriate Knowledge and Skills

Bhattacharyya has provided a list of appropriate knowledge and skills which are required, in addition to the professional education and training, for information work and services in different organizations.

2.3.1 Knowledge of Systems-Approach

This is the most appropriate mode of thinking applicable to any aspect of information service systems including their environments, super systems, objectives, subsystems (components), coordinate systems, and the functional and structural relationships between and among components. The results obtained by its application are most useful in designing or planning, developing, organizing, operating, controlling, understanding, explaining, and evaluating an information service system and all its components.

2.3.2 Knowledge of the Philosophy of Information Science

The term 'Philosophy' here refers to the organized body of knowledge that answers the questions of why, what, and how of information science and systems. It should necessarily comprehend all the working concepts, postulates, principles, and the rules of procedure that unify all the major components of information science forming the foundation of information service systems. This type of knowledge will help to study primary information as a phenomenon to the extent useful for secondary information services.

2.3.3 Knowledge of Information as a Resource Phenomenon; Use of Information; Users of Information; Information Service Profession; Information Service Systems, Centres, and Programmes; Environments of Information Generation, Communication and Services; and the Skill of Applying this Knowledge in Developing Professional

Tools, Techniques, and Services, and in Information Resource and Service Management

Knowledge in these areas is effectively used to the following information work and services:

(a) Dissemination of news about research in progress;

(b) Preparation of indexes and other aids to primary information embedded in individual documents;

(c) Dissemination and retrieval of information about, comparatively current, micro documents, and making provision of access to those documents;

(d) Preparation of non-critical factual documents, e.g. state-of-the-art reports, space-profiles, time-profiles, process-profiles, product-profiles, etc.;

(e) Acquisition, organization, storage, dissemination, and retrieval of information about macro documents, service of macro documents, etc.;

(f) Reprographic services; and

(g) Translation services.

2.3.4 Knowledge of Communication Channels or Media, and the Sources of Information (reference, primary, secondary, etc.), and the Skills of Generating and Rendering information Services

This knowledge will help in designing and developing new information services for their different user groups working in different organizational environments.

2.3.5 Knowledge of the Professional Techniques, Tools, and Services

Bhattacharyya suggested to acquire knowledge about:

(a) the techniques for collection development;

(b) classification, cataloguing, indexing, storing;

(c) repackaging, i.e., abstracting, extracting, digesting, and analysis and consolidation;

(d) tools, such as manuals, codes, standards, control devices; and

(e) information services, i.e., reference service, dissemination service, retrieval service, etc. Information professionals also need to acquire skills of applying this knowledge in designing, developing, and improving the techniques and tools meant for information services.

2.3.6 Knowledge of the Use of Appropriate Technologies

Information professionals must acquire knowledge of such technologies as computer, telecommunication, and graphic technologies including reprographic technologies to the extent necessary and sufficient. They may further add skills of applying this knowledge in housekeeping operations and in information storage, dissemination, retrieval, and other information services.

2.3.7 Knowledge of Scientific Management of a Service Production System

The term 'Management' refers, in terms of operations, primarily to planning, designing, developing, organizing, operating, controlling, and evaluating. All of them involve decision-making by considering all possible alternatives. In specific terms they manifest as manpower management, financial management, material management, equipment management, process management, product management, etc. Further, information professionals must acquire skills of applying this knowledge in the management of information service systems, centres, and programmes.

2.3.8 Knowledge of Appropriate Research Methodologies Applicable to the area of Information Service Systems

It is essential to acquire knowledge of research methodologies in one's own discipline, but it is equally important for information professionals to acquire knowledge of research methods in those disciplines found useful in relation to information service systems. These include: General Systems Theory, Logic, Mathematics, Statistics, Operations Research, Linguistics, and Psychology. In addition, "Technical Report Writing" is now considered as part of research methodologies. Knowledge of these disciplines and the skills of applying these are considered essential for preparing products of information analysis and consolidation.

2.3.9 Knowledge of Methodology of Studying Subjects for Information Work and Service

The primary concern of information service professionals is the information work and service predominantly pertaining to a subject or a group of subjects. Their professional work calls for necessary and sufficient knowledge of some of the special facets of the subject/s they are concerned with. It may be mentioned that no professional can learn all the facets of all the major subjects before entering into his professional job. Therefore, the solution lies in his learning a methodology of studying subjects for information work and service, so that he can use his skills in his professional career whenever the need arises. To develop his skill of applying it, one has to choose a subject that one can handle comfortably. A subject of one's academic study satisfies this criterion. One has to carry out a project to develop this skill. The quality of this knowledge and skill can further be improved upon.

2.4 Knowledge

Seetharama has put forward another line of thinking for another set of information services, that is, preparation of information consolidation services and products. He has given a list of topics about new knowledge to be acquired by information professionals pertaining to these services.

2.4.1 Methods for Studying the Highways and Byways of Subjects

The library and information professionals, in addition to the knowledge of their own subject, must also acquire knowledge about at least that subject/s which they are dealing with in the organization. In other words, as they cannot learn every aspect of all the subjects what they need to know are the methods for studying the highways and byways of subjects.

2.4.2 Data/Information Analysis Techniques

Adequate knowledge about the techniques and tools of information analysis is required. This includes sufficient knowledge about such techniques as indexing for the purpose of information/subject analysis, and its critical evaluation when working in an information centre. The techniques of indexing, abstracting, evaluation, etc. are part of the study in Library and Information Science courses where the information professionals get opportunity to acquire sufficient knowledge about these techniques.

2.4.3 Reference Framework Formulation

For designing an information consolidation service/ product, information workers have to determine the subject scope of the product. For this purpose, the information needs of the potential users are required to be studied. But in case of planners and decision-makers it is not so easy for their information needs may not be structured according to subjects. The use of Reference Framework approach is therefore resorted to, according to which, as suggested by Neelameghan and Seetharama, planning and decision-making programmes can be conceptually analyzed into different phases, sub-tasks, activities, etc. After such an analysis, it would be possible to select/extract relevant data and information from different sources/databases which can be repackaged into different subjects.

2.4.4 Methods for Planning and Conducting User Studies

There are various methods of collecting information about the requirements of potential users as well as about the organization in which they work. Information service professionals must acquire knowledge about these methods, which they can well acquire during their studies, and improve further on the job. These methods include:

- Interviewing users;
- Study of papers, books, etc. published by users;
- Attending seminars, symposia, colloquia, etc. in which users participate;
- Observing users at their workplace;
- Meeting users in small, preferably, homogeneous groups periodically;

– Personal information contact with users individually;
– Feedback from information services rendered; etc.

Since no single method may provide all the necessary information, therefore, several of the methods have to be used and the findings synthesized.

Saracevic and Wood have suggested that a user studies plan should consist of the following steps:

(a) Surveying the previous studies and literature in general;
(b) Determining the objectives of the study;
(c) Determining the variables to be studied;
(d) Selecting the sample of the population;
(e) Determining the method for collection of data;
(f) Determining the method of data/information analysis;
(g) Determining the ways of presentation and utilization of results.

2.4.5 Methods for Identification and Selection of Core Documents

Adequate knowledge about the identification and selection of core documents required for preparing an information consolidation product/service is quite significant. The library and information professionals know that one of the functions of the information centres is to select and acquire materials; therefore, they generally acquire such knowledge during their library and information science studies time. For the purpose of designing an ICP, they are required to select relevant documents keeping in view the interests of the users. They would select significant material like published from reputable sources, documents of foreign origin, primary periodicals, etc.

2.4.6 Preparation of Different Types of Information Consolidation Services/Products

Those involved in the preparation of information consolidation services/products must acquire adequate and sufficient knowledge about the preparation process. It involves the arrangement of ideas in a helpful sequence (say, according to S.R. Ranganathan's Principles of Helpful Sequence), and then integrating and consolidating the information collected into an organized text. The integration of information can be achieved by (a) application of classificatory principles; (b) adopting different methods of presentation of information; (c) technical writing; etc.

2.4.7 Techniques for Directory Preparation

The techniques for directory preparation are about the same as those for preparing any other Information Consolidation Product (ICP). However, those involved in the preparation of a directory of either the organization or the manpower working in an organization, must study the subject or the mission in which the information is to be analyzed. They must also have knowledge about the users, the objectives, resources, limitations, etc. They must also equip themselves with the specific guidelines on these areas and start work on it.

2.4.8 Condensation Techniques

Library and information professionals should acquire sufficient knowledge about the various condensation techniques, such as abstracting, extracting, etc. During their study period they have learnt about the techniques of preparing abstracts depending on their types. This knowledge needs to be improved upon further by familiarizing themselves with the latest guidelines for the purpose. They must also acquire necessary skills to apply these techniques while working in the information centres.

2.4.9 Evaluation Techniques for Assessing the Quality of ICPs

The users of an ICP usually tend to evaluate it, inter alia, against quality. The quality, according to Lancaster, would generally depend upon such considerations as the coverage, completeness or output (recall), relevance of output (precision), novelty of output, and completeness and accuracy of data. This describes well for the information service workers the need to acquaint themselves with the knowledge of the evaluation techniques to assess intrinsic merit of the ICPs.

2.4.10 Consolidation and Integration Techniques

Knowledge about the techniques of information consolidation and integration is essentially to be acquired by the library and information professionals while either in the library science schools or on the job. This knowledge also includes sufficient familiarity about the packaging, i.e. about package media, and package formats; and about repackaging of information which is rearrangement of physical media and/ or forms in which information is to be presented according to the requirements of users.

2.4.11 Guidelines for Arrangement of Ideas

The implication is that one and the same set of ideas can be arranged in several ways. The choice of a particular sequence should be based on its helpfulness to the users for whom the ICP is to be prepared. Library and information professionals acquire knowledge about the principles of helpful sequence, other classificatory principles during their study time. It can further be improved upon while working in information centre and preparing an ICP. They are required to know about the guidelines as to how the ideas are arranged in Idea Plane, Verbal Plane, and Notational Plane.

2.4.12 Technical Writing Techniques

While developing knowledge about research methods not only in Library and Information Science but in other subject/s also, the information professionals learn also about the technical writing techniques. These days technical writing is considered as a part of research methodologies where researchers are required to know as to how to write a technical report on their subjects. This knowledge is essential because information professionals have to prepare and write for ICPs from time to time.

2.4.13 Information Presentation Techniques

Information when presented systematically affects the stimulation, knowledge, current awareness, productivity, decision-making, problem-solving, etc. of the user recipients. Presentation of information therefore, assumes great significance in this context. Library and information professionals are required to determine the media—print media, audiovisual media, electronic media—and format in which Information Consolidation Product should be presented to enhance its potential use. They should also know the various presentation techniques like textual and non-textual, their advantages and limitations, etc.

2.4.14 Feedback Techniques

In order to increase interactive communication between the user groups and the information centre, there needs to set up a feedback mechanism. It will throw light on the lacunae and shortcomings, if any, in the ICPs and the information centre. The feedback received will help greatly in initiating necessary changes in the ICPs or the centre. Library and information professionals are required to know about the various techniques of obtaining feedback from different user groups. These may include personal contact, group personal contact, telephone, e-mail, website, etc.

3. Skills

The various areas of knowledge as mentioned above are the backbone for the information professionals engaged in information work and rendering information services, particularly the preparation of information products in the parent organizations. Some components of this knowledge have been acquired by the professionals during their study in library and information schools, while others may be updated while working on the job.

It is, however, argued successfully that mere acquisition of knowledge about various facets of library and information science may not be sufficient for rendering specialized information services as well as for preparing ICPs. The information professionals must also possess adequate skills to enable them to apply knowledge gained fruitfully. Seetharama, therefore, suggests that it is necessary for information professionals to possess the following skills:

3.1 Traditional Skills

Information service professionals, as mentioned above, have acquired knowledge about the traditional techniques of collection development, organizing, i.e., classification, cataloguing, indexing, etc. They now have to develop the skills of acquiring, organizing, and dissemination of information. By applying the traditional skills information professionals can design, develop, and improve dissemination process as well as information consolidation products for their user groups.

3.2 Skills in Communication, etc.

It is important for information professionals in the present times to possess adequate skills in communication, systems analysis, interpersonal relations, financial planning, and marketing. Skills in communication are necessary in transferring and disseminating various types of information about services as well as products.

By applying skills of systems analysis, helpful results are obtained in planning, designing, developing, etc. of consolidated information services and products.

In view of paucity of adequate funds skills in financial planning become quite handy for the information professionals.

Similarly, skills in marketing would go a long way in not only studying the potential market for the information services and products but would also open the highways and byways for their promotion and dissemination.

3.3 Managerial Skills for Technology Management

Acquisition and application of managerial skills are required much more while dealing with the technology and its application to library and information work, like house keeping operations, information storage, dissemination, etc. Adequate skills in problem-solving and decision-making, etc. are desired not only to take difficult decisions in certain situations but also for problem-solving while designing and developing consolidated information services and products.

3.4 Skills to Analyze, Synthesize, etc.

Preparation of Information Consolidation services and products depends upon information analysis, synthesis, interpretation, and evaluation, etc. Therefore, the skills regarding subject analysis, synthesizing information in a coherent manner, evaluation of consolidated information against quality, its packaging and repackaging in different physical media and/or forms in which it has to be presented, have not only to be acquired but improved upon regularly for efficiency and effectiveness of services and products.

3.5 Advocacy Skills

According to Bearman, it is significant for information professionals to develop advocacy skills with a greater sensitivity to social and political forces and an increased ability to define and articulate value and value systems. The role of advocate, according to Toni Carbo, is becoming increasingly important—whether it is in the protection of civil liberties, the library budget, or the services required to meet users' needs.

4. Other Skills

When the information professionals are able to acquire and develop the above-mentioned knowledge and skills they are in a position to "recognize and assess the implications of their actions and actions of others". In this regard Lynda Resnik, an American entrepreneur and business woman, points out that such skills must be planted and nurtured and not just left to chance. She has observed that for the preparation of information consolidated products the following skills and/or abilities would be required:

1. To interact with users and recognize their informational needs.
2. To identify and organize a team or group of specialists for the preparation of information consolidation products.
3. To specify precisely the subject to be covered by the information consolidation product.
4. To specify the product user group and target group.

5. To identify the products to be prepared for different user groups.
6. To identify sources for selection of information.
7. To identify and select relevant information.
8. To technically process and evaluate information.
9. To consolidate and integrate information into an organized text.
10. To prepare and provide supplemental aids to the users of information consolidation products.

4.1 Competencies

In this information age, knowledgeable, skilled and competent library and informational professionals understand the critical role that information plays for organizations and for individuals. In order to fulfill this key information role, they require two main types of competencies:

4.1.1 Professional Competencies

These relate to library and information professional's knowledge in the areas of information resources, information, access, technology, management, etc. for providing library and information services.

4.1.2 Personal Competencies

These represent a set of skills, attitudes and values that enable library and information professionals to work efficiently; be good communicators; focus on continuing learning; show value-added nature of their contributions, and so on.

5. Future Information Professionals

India is a society in the grip of rapid economic, social, and technological transformation which is reshaping the life of people. All aspects of human society, including higher education, are being impacted by these developments. In the present Information Society/Knowledge Society, the subject of Library and Information Science, which is concerned with information and knowledge organization as well as dissemination, is assuming greater importance. The information professionals of the 21st century have to equip themselves with the latest knowledge of handling techniques and skills of storage and retrieval of information. The knowledge and skills they acquire must keep pace with the changes required in the organization to improve its performance.

The library and information professionals can be prepared for tomorrow by reengineering the process of education with required vision. To quote Richard Budd, "We don't train them for their first job, we educate them for their last job". They must know the fundamental of the field of library and information science, the management and other skills as discussed above.

Many think tanks, such as Becker and others, assert that librarians and information professionals are destined to play increasing roles in establishing connections between information resources and people in the Information Society. While agreeing with this line of thinking, Foskett commented that "...it would be a tragic irony if the medium becomes the message, technology becomes the master and not the servant and librarians found themselves once more cast in the role of Keepers, not of books, but of machines!... Librarians must not neglect their role as repositories, but become also mediators and promoters of communication...."

For developing and sharpening communication skills, information professionals need to keep inventing "bridging the gap" technologies, and "keep discovering the language of the user".

6. Manpower Development

Like other professions, library and information profession also has a manpower development system. The ultimate aim of such a system is developing manpower for information work and services in various organizations. They have to be equipped with appropriate knowledge and skills, as discussed above. This knowledge and skills are acquired through professional education and training imparted in LIS schools/departments. In these schools/departments every person entering with this purpose is enriched with appropriate knowledge and skills and developed into appropriate manpower for information work and services in different institutional environments.

At present in India, almost all the LIS schools/departments are offering either a full paper or as a part of another related paper on Information Analysis and Consolidation. Besides, the National Information on Science Communication and Information Resources (NISCAIR), New Delhi and the Documentation Research and Training Centre (DRTC), Bangalore, offer a full-time post graduate Two-year course in Library and Information Science with adequate emphasis on Information Consolidation. These two centres also offer short-term courses and on-the-job training in Information Consolidation from time to time for updating the knowledge and skills of working library and information professionals.

Such continuing education programmes need to be organized frequently at different levels in different parts of the country so that most information workers are adequately equipped in information analysis and consolidation skills. They must, as remarked by Toni Carbo, show curiosity and willingness to continue learning, i.e. "learning to learn" because it is important in this context.

In order to update their knowledge and skills further, they may be provided adequate opportunities for participating in national and international seminars, conferences, symposia, etc. particularly those related to information consolidation activities. Seetharama has suggested that the staff working in the Information Consolidation Units should be encouraged to "take up projects involving research, survey, and in-depth study relating to various problems arising form their information activities." This and other efforts will make the information professionals innovate their services to keep ahead of users' needs to enrich users' knowledge about their research projects and other information needs.

References

Agada, John. "Information Technology and the Preparation of Librarians for the Information Society: Questions for Developing Countries". *Personnel Training and Education,* Vol. 7 (3) 1990.

Atherton, P. *Handbook of Information Systems and Services.* Paris: UNESCO, 1977.

Bearman, T.C. "Educating the Future Information Professional". *Library Hitech Issue,* Vol. 18, 1984: 27-40.

Bhattacharyya, G. "Knowledge and Skills for Secondary Information Work and Service: Foundation of Curricula and Syllabi". Tejomurty, A., Ed.: *Restructuring of Library and Information Science Curriculum.* New Delhi: Concept Publishing Company, 1994, pp. 63-74.

Carbo, Toni. "The Future of Librarianship: A View from a School of Library and Information Science." *Library Connect Newsletter*, Vol. 1(4), December 2003: 5-6.

International Federation of Library Associations and Institutions (IFLA). *Guidelines for Professional Library/Information Education Programs.* Latest revision 2009. (http://www.efla.org/VII/s23/bulletin/guidelines.htm).

Kawatra, P.S., Ed. *World Encyclopedia of Library and Information Science Education.* 3 Volumes. New Delhi: Crest Publishing House, 1994.

Lancaster, F.W. *Guidelines for the Preparation of Training Courses, Workshops and Seminars.* Paris: UNESCO, Bureau of Studies and Programming, 1983.

Laxman Rao, N. et al. Eds. *Changing Dimensions of LIS Education: Seminar Papers.* XVIII IATLIS National Seminar, Andhra University, Visakhapatnam, 12-14 October, 2001. Hyderabad: IATLIS, 2001.

Naylor, R.J. "Core Competencies: What They are and How to use them". *Public Libraries*, 39, 2000: 108-114.

Neelameghan, A. and Seetharama, S. "Reference Framework: An Aid to Information Analysis and Consolidation." In *DRTC Annual Seminar*, 18, 1981, Paper BD.

Parekh, Harsha. *Bridging the Divide: Perspectives of an Information Professional.* Compiled by Bharati Sen and Nila Jani. Mumbai: SHPT School of Library Science, 2003.

Resnik, Lynda. *Rubies in the Orchard: How to Uncover the Hidden Gems in Your Business.* Crown Publishing, 2009.

Saracevic, T. *A Course in Information Consolidation: A Handbook for Education and Training in Analysis, Synthesis and Repackaging of Information (Preliminary Version).* Paris: UNESCO, GIP and UNISIST, 1986.

Saracevic, T. and Wood, J.B. *Consolidation of Information: A Handbook of Evaluation, Restructuring and Repackaging of Scientific and Technical Information.* Paris: UNESCO, 1981.

——. *Information Consolidation and Repackaging: Framework, Methodology, Planning.* New Delhi: Ess Ess Publications, 1997.

Seetharama, S. "Information Professionals and their Skills for Future Information Society". Paper Presented at *FID-88 Conference,* Helsinki, Finland, August 1988.

Special Libraries Association. *Competencies for Special Librarians of the 21st Century: Full Report*, submitted by Joanne Marshall, et al. Washington: SLA, 1996. Revised June 2003.

Evaluation of Information Products

14

1. Introduction

Evaluation is concerned with determination of the intrinsic merit, validity and reliability of information products and services. In other words, it is the quality of information products and services that concerns the users. The quality of information services and products depends much on the quality of information collected, analyzed and synthesized for the purpose of preparing information products. After preparing and distributing the information consolidated products they need to be evaluated on a regular basis. Their evaluation needs to be based on users' criteria of quality, completeness, comprehensiveness, appropriateness, time lag, use, costs, etc. In other words, evaluation of information products and services should not be considered merely a one-time activity. In order to improve the effectiveness and performance of the products evaluation is required to be done on a continual basis, primarily from the point of view of the users of information services and products.

2. Benefits

Some of the benefits of evaluation of information/ information services/information products may be as follows, though it may not be easy to pinpoint them, and these may also not be easy to quantify them.

(i) increased productivity

(ii) increased well-being

(iii) better life

(iv) comprehensive understanding of a problem

(v) more certainty in decision-making

(vi) easier actions

(vii) promotion

(viii) healthier surroundings, etc.

3. Criteria

Many studies have been performed on the value of information services and products to the users in their projects, research, organizational work, etc. These studies have been made by scholars like Robert Mason, Lancaster, etc. about the products of information systems in general and information analysis centres in particular. These studies have attempted to define a set of criteria by which users tend to evaluate and judge information services and products. Lancaster has suggested the following criteria deemed to be most important for evaluation:

4. Evaluation of Effectiveness

Evaluation of effectiveness of information products and services concerns primarily to the users working in different organizations. Their considerations about the effectiveness are as follows:

4.1 Cost Criteria

The cost of information products plays an important role in judging their effectiveness from users' point of view. The following factors need to be given due consideration in this regard.

(i) Monetary Cost to User—It is measured in terms of cost per search made from the online or offline database, by the user in response to a query. This cost is also measured on subscription of a product or service from an organization. It also costs the users when they buy a document for meeting the requirements of their research projects.

(ii) Less Tangible Cost Considerations—In addition to tangible costs, there are some less tangible costs, but these are equally important. These include:

(a) Effort involved in learning how to use the system;

(b) Effort involved in actual use of the system;

(c) Effort involved in retrieving documents (through backup document delivery system); and

(d) Form of output provided by the system.

4.2 Time Criteria

Timeliness of an information product is very significant because information contained in the product looses its value if time lag occurs beyond admissible level. It is explained in the following situations:

(i) Time elapsing from submission of request to retrieval of citations;

(ii) Time elapsing from submission of request to retrieval of documents;

(iii) Other time considerations, for example, waiting time to use an online system.

4.3 Quality Considerations

These are mentioned as follows:

(i) Coverage: The coverage of information product from the point of view of comprehensiveness of subject or topics has significant value for quality.

(ii) Completeness of Output (Recall): This is another consideration describing quality of the product and service. An information product prepared with completeness of information shall have the maximum recall values.

(iii) Relevance of Output (Precision): The quality of the information product will be judged by the users against the precision of information when recalled in response to a query.

(iv) Novelty of Output: A qualitative information product must be able to provide new and unique information to users of the product.

(v) Completeness and Accuracy of Data: The value of complete and accurate data or/and information need not be overstated. Completeness and accuracy of data/information included in the information product speaks volumes about its quality from users' point of view.

(vi) Credibility of the Sources: The sources from which the required information has been collected must be dependable enough from the users' point of view.

(vii) Recency of Information: It goes without saying that the information or data to be used must be recent to provide the up to date account about the subject.

5. Evaluation of Cost Effectiveness

It is concerned with the user satisfaction related to internal efficiency and cost considerations. It takes into account the following factors:

(i) Unit Cost per Relevant Citation Retrieved: Unit cost has to be measured in terms of relevant citations retrieved from the database. The relevant citations describe the effectiveness of the retrieved citations and their cost need to be worked out accordingly.

(ii) Unit Cost per New Relevant Citation Retrieved: While retrieving citations from the database developed for the information product and/or service, the cost of only those relevant citations is to be worked out which are new, that is, previously unknown to the experts.

(iii) Unit Cost per Relevant Document Retrieved: Valuation of cost effectiveness of information product can be measured in terms of unit cost per relevant document retrieved from the database.

6. Cost-Benefit Evaluation

It covers the value of system balanced against costs of operating it. In this context, Vickery's observation on Ranganathan's Five Laws of Library Science, especially the Fourth Law, is very relevant. He remarks ".... The principle save the time of the reader and staff is now most actively pursued within documentation, in such guises as the evaluation of the performance of services, cost-effectiveness and cost benefit. In stressing 'time', Ranganathan pointed to an element of system performance which is often of crucial subjective importance to the user (and indeed the author); the time involved in getting documents published, distributed, processed, located, delivered; the 'timeliness' with which a document reaches a user. There is also the trade-off between the time of the user and the time of the staff—a faster service may demand more staff...." As a matter of fact, all the Five Laws are very good normative principles and can be used as a basis for deriving evaluation criteria of the Information Consolidation Unit (ICU) and its products and services.

Seetharama remarks that a similar line of thinking can be seen in the suggestion that the evaluation of an Information Consolidation Unit should be based on:

(a) Achievement of objectives
(b) Economic aspects (costs)
(c) Feedback from users (have needs been met, levels of user satisfaction/penetration achieved)
(d) Efficiency and effectiveness of performance
(e) Actual benefits derived (social and economic benefits as well)
(f) Effects on role played in advancing technology

It has, however, been pointed out by Saracevic and Wood that in different situations different priorities may be assigned to each of these dimensions. In addition, other specific criteria may be used by different users in the same situation and by the same user in different situations.

7. Evaluation Steps

Evaluation of information products and services requires some steps to be taken by the evaluators. Lancaster has suggested the following major steps for an evaluation programme. These steps are also relevant to the evaluation of information products and services of an ICU.

(i) Defining the scope of evaluation
(ii) Designing the evaluation programme

(iii) Execution of the evaluation
(iv) Analysis and interpretation of results
(v) Modifying the system or service on the basis of evaluation results

However, Brandreth has provided a note of caution, especially when evaluator is working for gathering requisite data/information. He says that too zealous data-gathering could rather impede operations and consume too many resources of the system. He calls it as *a paralysis by analysis*.

Another important point is that failures in ICUs should be as thoroughly analyzed as successes. This is what Saracevic and Wood remarked that information consolidation requires not only a study of options and successes but also of failures. Some factors, identified by Saracevic and Wood, as contributory to the failure of information consolidation products are as follows:

(a) No market for information product or services,
(b) Market is dynamic, product or service is not,
(c) Products and/or services are usually designed for information specialists or librarians, not for users,
(d) Some products and/or services available elsewhere under better conditions,
(e) Difficult to obtain, access, and use,
(f) Price is too high, or hidden costs in use are too high, or too time consuming in use,
(g) Lack of direct link with information producing institution,
(h) Inadequate staff, or too small a staff,
(i) Inadequate economic and technical support and resources, and so on.

8. Who Should Evaluate

The big questions are: who should be the evaluators?, and how should they go about doing the evaluation? Saracevic and Wood believe that some degree of subject expertise is necessary for a person to be an evaluator. The more specialized the subject and the clientele the greater the expertise required. But studies show that even the experts tend to disagree in their evaluations. The only way to resolve this problem of evaluation is to search for repeated evaluations and for some sort of consensus.

However, the ideal situation is to have evaluators that are both: experts in a subject and experts in information services (including users). But the problem arises when a subject expert also assumes to be an expert in information or library expert. Unfortunately, the ideal situation is rarely achieved.

Kertesz has suggested that evaluation should be done by an Internal Review Committee as well as by an outside Expert Group. The Internal Review Committee will be involved in the evaluation of the routine operations of the ICU and should consist of members of the host institutions/sponsoring agency, supplemented by outside experts, if desired. This Committee is expected to maintain a closer contact with operations; its periodic reports should be primarily for the benefit of top management and the sponsors to set policy and improve services.

On the other hand, the outside Expert Group forming the Expert Review Committee should evaluate the quality of the output and of the services rendered by the ICU. Their role should be strictly advisory, without assuming any management function.

References

Brandreth, M. *Specialised Information/Analysis Centres in International Development: Report of a meeting* held at Montebellow, Quebec, Canada, 4-8 October, 1982. IRDC Manuscript Report Series. March 1984.

Kertesz, F. "Collaboration between IACs at a Large Multipurpose Laboratory", Quoted in Weisman, H.M. *Information Systems, Services and Centres*. New York: Becker and Hayes, 1972.

Lancaster, F.W. *Guidelines for the Evaluation of Information Systems and Services*. Paris: UNESCO, 1978.

——. *If You Want to Evaluate Your Library*. 2nd ed. Champaign, IL: University of Ilinoid, 1993.

——. *Information Retrieval Systems: Characteristics, Testing and Evaluation*. 2nd ed. New York: John Wiley, 1979.

——. *The Measurement and Evaluation of Library Services*. 2nd ed. Arlington, Va: Information Resources Press, 1991.

Saracevic, T. *A Course in Information Consolidation: A Handbook for Education and Training in Analysis, Synthesis and Repackaging of Information* (Preliminary version). Paris: UNESCO, 1986.

——. "Processes and Problems in Information Consolidation." *Information Processing and Management*, Volume 22(1) 1986: 45-60.

Saracevic, T. and Wood, J.B. *Consolidation of Information: A Handbook on Evaluation, Restructuring and Repackaging of Scientific and Technical Information*. Paris: UNESCO, 1981.

Seetharama, S. *Information Consolidation and Repackaging: Framework, Methodology, Planning*. New Delhi: Ess Ess Publications, 1997.

Vickery, B.C. "Concepts of Documentation". *Journal of Documentation*, Vol. 34, 1978: 279-287.

Weisman, Herman M. *Information Systems, Services, and Centres*. New York: Becker and Hayes, 1972.

Trends in Information Analysis and Consolidation

15

1. Introduction

The present society is characterized by continuous change as a result of which it has evolved into what is termed as Information Society. Libraries and information centres have also been changing form simple to complex organizations due to the profound impact of information technology (IT). Information professionals have been feeling not only to maximize the use of available information sources, but also to develop new information products and services. This has been facilitated by the available IT and other communication technologies.

Information professionals equipped with the appropriate knowledge and skills, as already discussed, will be able to analyze, synthesize, consolidate, and repackage relevant information into meaningful services and products. Such consolidated information services and products will be made available to the right users at the right time in the right format to meet their needs.

In this fast changing scenario, the latest trends in information analysis, consolidation, packaging and repackaging are visible when the new formats of electronic content creation and its use in electronic databases are there to stay.

2. Information Analysis

Seetharama observes that the term "Information Analysis" in the context of various activities in the field of information does not connote and denote one and the same idea operationally. While in the context of indexing it is subject analysis, it is considered to mean critical evaluation when one refers to an information analysis centre. In the spectrum delimited by these two extreme points, information analysis is being used to denote a very specific species of activity in the field of secondary information services especially when it goes with the term "consolidation". In the context of information consolidation, it is used primarily to denote a non-evaluative information analysis for the purpose of consolidating the results of analysis into a textual presentation in which the isolated units of input undergo substantial structural changes. Thus, the product is always factual, non-critical, or non-evaluative, but informative in nature. It is a firm, coherent, and unified body of information.

This approach is slightly different from other descriptions wherein it is emphasized that the importance of consolidation function lies in evaluation and it is this function that distinguishes information consolidation from information analysis and other processes such as indexing, abstracting, etc.

3. Information Consolidation

In Unesco Symposium on Information Analysis and Consolidation held in Sri Lanka in 1978, information consolidation has been defined as follows:

> Information Consolidation Activities are used to define the responsibility exercised by individuals, departments or organizations for evaluating and compressing relevant documents in order to provide definite user groups with reliable and concise new bodies of knowledge. Individuals or groups of individuals performing consolidation activities would each constitute an Information Consolidation Unit.

Saracevic and Wood have put forward the following comprehensive and concise definition of information consolidation. They said:

> Consolidated Information is public knowledge specifically selected, analysed, and possibly

restructured and repackaged for the purpose of serving some of the immediate decisions, problems, and information needs of a defined clientele or social groups, who otherwise may not be able to effectively and efficiently access and use this knowledge as available in the great amounts of documents or its original form. The criteria for selection, evaluation, restructuring, and repackaging of this knowledge are derived from the potential clientele.

In other words, "Consolidated Information is a text(s) or message(s) purposefully structured from existing public knowledge to affect the private knowledge and decisions of individuals who otherwise may not be able to effectively access or use this public knowledge from the original amounts or in the original structure and form."

4. Information Packaging

Packaging of information is a physical recording, arrangement and presentation of information on a given medium and in a given form. According to Saracevic and Wood, two aspects can be distinguished in packaging as follows:

4.1. Package Media: The reference here is to physical medium in which the information is displayed or presented to users. For example, the physical media can be print, audio-visual, demonstrations, interpersonal contact, etc.

4.2. Package Formats: This refers to the arrangement, shape, and layout of information in a given production on a given medium.

It may be noted that while packaging cannot improve information content, it can enhance its value by attractive looks and thus lead to increased use, assimilation and recall.

5. Information Repackaging

Saracevic and Wood stated that repackaging of information is rearrangement of physical media and/or forms in which information has been presented. It also refers to presentation in a given medium and form of restructured information. Repackaging is also carried out to make the consolidated product tailored to the requirements of a specific group of users. In other words, repackaging of information refers to the presentation of information in more understandable, readable, acceptable and usable forms.

The aim of repackaging is to enhance the acceptance and use of information products and the assimilation and recall of their contents.

6. Trends in Information Analysis, Consolidation and Repackaging

It is generally held that information analysis and consolidation activities are not of recent origin; its history, however, can be traced back to the nineteenth century. It was developed to cope with the challenges of the ever-increasing flood of scientific information and data and to make experimental findings conveniently available to the scientists and researchers, etc. by bringing some order in a usable form for the users in the form of some kind of information product.

However, the term "Information Analysis and Consolidation" is rather recent. Broadly speaking, information analysis and consolidation activities include selection, evaluation, analysis, interpretation and synthesis of a body of information in a specialized field. The purpose is compiling, digesting, repackaging and presenting pertinent information in an acceptable form. The information should be timely and useful to meet with the different requirements of different user groups.

The origin of the Information Analysis Centres (IAC), also referred to as Information Consolidation Units (ICU), was due to the untiring efforts of a few dedicated specialists, particularly in scientific disciplines. On the other hand, the present day ICUs have largely been due to the efforts of various sponsoring agencies belonging to both the private and public sectors. Thus, presently a need has arisen for justifying the information consolidation activities to the satisfaction of all concerned including various user groups.

7. Benefits of Use of Consolidated Information

The current trends in applying information analysis, consolidation and repackaging reveal that there are several benefits for the information users as well as the information professionals.

7.1. ***Improved Decision-making***—Use of consolidated information helps users to be better informed about alternatives and consequences. This leads to improved decision-making and reduces uncertainty in the process.

7.2. ***Knowledge Competence***—For users making use of consolidated information, it helps to increase level, depth and breadth of knowledge by individual users and users in groups. This also leads to higher sophistication in drawing relations between seemingly unrelated facts.

7.3. ***Adaptation***—Consolidated information helps users to have more appropriate and adjusted responses by individuals, groups or organizations to demands and changes in the environment and a complex world.

7.4. ***Productivity***—Organizations carrying out information consolidation activities have higher levels of outputs in work and other activities. As a result, the capacity to deliver products too increases with better effectiveness. It also adds more contacts in favour of such organizations.

7.5. ***Resources***—Information consolidation activities lead to efficient utilization of limited resources of the information organizations. It ensures increased capacity and effectiveness in its performance. The end result is economy in every aspect of organizational resources.

7.6. ***Success***—Better success is ensured due to contribution towards attaining objectives as laid down by individual users, groups, organizations, and all concerned. Use of consolidated information will give a wider spread and acceptance of results. It is also possible to detect and make necessary adjustments in the aims and objectives.

7.7. ***Business Enterprises***—Information consolidation products help the users involved in business enterprises, commerce, and related areas. They provide appropriate responses to market conditions and demands. The use of these products enables the users meeting competition, complying with standards and regulations. It also motivates personnel and helps to make business decisions in their favour.

7.8. Manufacturing Industry—In manufacturing industry information consolidation products help incorporate much needed technical innovations. Their use enables the industry not only adding new products, but also making their products more appropriate, reducing breakdowns, and increasing productivity in industry.

7.9. Science—In scientific areas, the information consolidation products help to keep up with research front by assisting users to make appropriate decisions on policy and allocations issues. The users can make judgments regarding their own position and the advances they have made in this regard. They can also search for related works and/or ideas for further work and methodology.

7.10. Education—The users in educational field can keep up with advances made in given fields as well as in education, research, methods, approaches, etc. The use of information consolidation products also helps the users in educational planning and assessments or comparisons.

7.11. Individuals—For individual users consolidated information provide for self-fulfillment and advancement in their given areas of work or interest. Information consolidated products in various specialized subjects also provide for wider opportunities in employment, self help and adjustments to changing environments and conditions. Such a development leads to increased sophistication towards higher quality of life and work situations.

Saracevic and Wood believe that information consolidated products can play a significant role because the process is an attempt to make available and accessible to users information that is appropriate to the tasks at hand, and structured and packaged to coincide with their level of understanding.

They further said that information consolidation can also be viewed as an intellectual post-processing of the available primary and secondary information to make it accessible to users, who otherwise would not be able to access this information. At the same time it can be recognized that information consolidation is just one in the spectrum of activities relevant to bring about an increase in information use for development. This is because of the fact that specific types of information consolidation are tied to specific user demands and other specific information services.

Thus, it can be concluded that consolidated information is essential in the framework necessary for bringing social change and development.

8. Electronic Content Creation

It seems likely that printed documentation will remain necessary in many areas of technical publications, which include information analysis and consolidation products. But in the near future, documentation for the computer industry is moving ever closer to a paperless concept, particularly with the growth of multimedia systems that integrate video and sound. Online documentation, because of its very nature, requires a screen on which to display and this restricts it for the most part to computer application only.

Help text for the software packages is the most common form of online documentation. Others include training and tutorial application, onscreen demonstrations, online access databases, CD-ROM based databases and animated displays, hypertext, creation of web pages for websites, video and sound.

8.1 Help Systems

Help text systems for software applications comprise full or part screen displays, usually retrieved by a single keystroke, such as a function key, or a key combination. Text must be displayed to provide an overall guide to the application—almost like an onscreen manual—through which the user can browse to find relevant information. Alternatively, it may also be restricted to covering a single topic relevant to the process being performed. This latter case is known as 'context sensitive' help system.

Numerous and varied facilities are available for creating online documentation as they are so often developed as part of the application they are intended to support. The main exception to this is in industry standard operating environments, like Microsoft Windows, in which all applications tend to provide help text that operates in the same manner and the tools for editing and generating the help information can be applied to any application working in that operating environment.

Help systems require a structured approach to providing information and provide an added challenge to the authors in that they often have to communicate effectively in as few words as possible. This could be argued as the same goal for those writing/printing documentation, the help text author may not have the benefit of using the typographical or illustrative elements on a small screen displays as can be provided on a printed page. Nevertheless, the rapid development of more powerful and efficient personal computers and processors means that the tools available to those involved in online documentation will be ever more effective.

8.2 Hyper Text

Hyper Text systems provide a mixture of online help and tutorial information to enable users to work in an interactive interface. This means that the operator can choose to move around help system on a key work or topic basis in order to find out more about a particular piece of information. Some help systems too come close to be hypertext themselves. According to Rowley, Hyper Text Mark up Language (HTML) allows browsers on the Web to see formatted text. Any text marked up with HTML can be viewed on any browsing system to the full capacity of that system.

In the Microsoft Windows operating environment, context sensitive help can be compiled with what are called 'hot keys'. These are words or phrases or even graphic that can act as a trigger to retrieve more information about a related topic. These keywords or phrased that form the hot keys may be displayed in a different colour so that the user knows that by clicking on a hot key with the mouse pointer another window of help information may be displayed offering another level of details about the topic concerned. The words with dashed underlining are the hot keywords and phrases.

8.3 Expert Systems

Expert system (ES) is the result of the most practical application of artificial intelligence. Expert system is a computer programme designed to reason like a human expert in a given situation. There are some software systems that offer this form of online documentation in association with an application. Rowley states that knowledge is often represented in an ES in a knowledge base, which is a network of interconnected rules which represent the human expertise. ES are a type of decision support system, hence can solve the problems in a way similar to that of a human expert and communicate the decisions with a very fast speed. Essentially, it is a database of information that provides answers to questions posed by the users. Those questions are inevitably structured and in some cases the user may be given a limited choice of options. In response to the users' questions, some solution, action, or simply more information can be provided by using a structured and logical design. The user gradually arrives at the information or action required by responding to progressively more detailed queries. This makes Expert Systems ideally suited to handling troubleshooting documentation. Expert Systems, like hypertext system of help, can be used in place of or alongside traditional documentation. The development of such systems has been going on for quite some time but they still have to realize their full promised potential.

8.4 Multimedia Systems

Multimedia covers a significant range of additional specialist skills through animation, graphics, video imaging and processing, sound recording and mixing as well as data and text-based authorship, which play a prominent role in the technical documentation. In recent times, this form of communication has become quite important as the need for printed documentation is gradually eroding in the computing industry. However, printed documentation remains an important component in this field at present as well as being an essential accompaniment to hardware and systems that cannot provide a means of interactive communication through the product itself. The role of the technical author is likely to change in the future as it will become more specialized and as printed matter is generally being replaced by multimedia forms of communication. Rowley observes that although it is difficult to handle non-text items such as drawings and moving images, but these can be indexed with keywords in a similar manner to text-based documents for sophisticated access.

8.5 Website Preparation

A web page is just like a file, an HTML file to be exact. It is called HTML because webpage documents have the file extension.html or.htm. HTML stands for Hyper Text Mark-up Language. The web is becoming an integral part of our working world. In a very short-time span, the web has revolutionized the way we access information, education, business, entertainment, etc. It has created industries where there were none before. Being able to develop information on the web might be a job skill, a class requirement, a business necessity, or a personal interest. Unlike any other previously discussed medium, the ability to "write" HTML allows to potentially connect with millions of other people who may become a kind of self publisher for an individual user.

Web pages are actually text files that are "tagged" with symbols to represent structure and function. A user can prepare and edit Web pages in any text editor or word processor he/she likes. The system of tags, called HTML, is very logical and not difficult to learn. However, one need not be an expert in HTML in order to create own Web pages. There are several shortcuts that have been developed to simplify the process of creating it.

The icons, pictures, sound bites, and video clips that are common on Web page are actually separate files that are referred to by special HTML codes. When the page is viewed by a Web browser, such as Netscape or Internet Explorer, these HTML codes are interpreted by the browser, which pulls in all the separate files and assembles the Web page in the form one finally views it. This is in contrast to a modern word processor, in which the editing and viewing functions are integrated into one program and in which all the text, graphics and other elements of the document are contained in one document file. For that reason, the process of creating Web pages is less direct than creating an equivalent word processor document.

8.6 Desktop Publishing

It is a combination of editing and word processing. As an editor the desktop publisher ensures that whatever is printed exhibits a high standard of professionalism. Desktop publishing packages allow the user to create documents including text and graphics. As a word processor the desktop publisher creates a flawless file which may be transmitted electronically and printed identically worldwide. The desktop publisher enables the user to set up the page and other document design for a document, so that it can be set up ready for printing. The role of desktop publisher thus is to ensure absolute document clarity and adherence to strict editorial layout standards. Many word-processing packages now offer some desktop publishing features.

Desktop publishing has major role to play in the printing process by providing the following services:

(a) To prepare and print the documents in-house;

(b) To prepare and print a camera-ready-copy of the document for publishing by a commercial printer;

(c) To oversee commercial printing, binding and document distribution;

(d) To prepare and electronically transmit the document files worldwide for local publishing; and

(e) To convert the document into another media.

9. Online Documentation

Documentation is important throughout the process of systems design as well as product design. It gives a clear picture of work completed and supports good communication between everyone concerned. Online documentation is information displayed, usually on a computer screen, which informs or instructs the users of a system or application. It may be presented in many forms.

Information analysis and consolidation products, also referred to as technical documents, are prepared for ensuring proper communication to the users. For carrying out this activity, criteria for technical writing are to be adhered to, and all good technical documents meet the following requirements:

(a) They contain complete information;

(b) They contain easy-to-find information;

(c) They are attractive in format;

(d) They should inform users how to use the information contained in it;

(e) They must contain pertinent information and well explained from the users' point of view.

Online software documents are web-based information products which allow the writers to use multiple typeface styles, sizes, coloured text, symbols, graphics, and hypertext links. These documents are usually prepared by a team of specialists which include subject experts, library and information professionals, technical writers, and a programming team. For the sake of accuracy, the software documents are reviewed, edited, tested, and approved by professional technical writers and subject experts. Once approved a notification is issued regarding the existence of new software documents or the one with changes to the already existing documents.

The objective of these documents is that the users find information quickly, or learn easily about the product or service in depth. For good understanding these documents should therefore be jargon-free and in non-technical language. These are delivered in various forms such as manuals, windows help, browser help, Adobe Acrobat, PDF documents, OCR, etc.

10. Sum up

The aim of information consolidation, as discussed above, is to increase the effectiveness of usage of information,

information services and products as well as to increase the circle of the users served.

It is said that information consolidation is neither an information panacea, nor a substitute for any of the existing information activities. It is just an approach in the whole spectrum of information practices needed for various aspects of development.

However, information consolidation is an effective approach to the fulfillment of specific information needs for evaluative and synthesized information services for the service of the yet unserved. Information consolidation of one type is applicable to highly sophisticated users, viz., scientists, engineers, managers, policy makers, etc., and of another type to users of lesser sophistication, viz., workers, villagers, school children, and the like. The choices are wide. They can be made only on the basis of priorities and value in specific developmental efforts and environments.

Information consolidation is complex and demands a definite commitment of resources, such as information, financial, manpower, physical, technical, etc. However, information consolidation has proved its value and benefits in information transfer and communication. Therefore, regardless of its complexity, information consolidation should be given a high consideration in the institution or refinement of information efforts serving development for social change.

It can be summed up with a word of advice from Saracevic and Wood regarding the secret of information consolidation which is this:

> ...while information consolidation involves complexity in processes, it offers simplicity and appropriateness in products. The potential benefits derived in using consolidated information are high.

To conclude it may be said that in the present era of information/knowledge society where electronic information is all pervasive, information consolidation and repackaging is a way of improving library and information services. Library and information professionals must take up the clarion call and implement information repackaging in the entire array of library services. This effort will go a long way in making the users information rich to serve the society better.

References

Atherton, Pauline. *Handbook for Information Systems and Services*. Paris: UNESCO, 1977.

Bhattacharyya, G. "Information Analysis for Consolidation: Some Basic Considerations". *DRTC Annual Seminar*. Vol. 18. 1981, Paper AA.

Documentation Research and Training Centre, Bangalore. *Information Analysis and Consolidation: Principles, Procedures and Products*, by G. Bhattacharyya and M.A. Gopinath. Annual Seminar. 18. Bangalore: DRTC, 1981.

Rowley, Jennifer. *The Basics of Information Systems*. 2nd ed. London: Library Association Publishing, 1996.

Saracevic, T. *A Course in Information Consolidation: A Handbook for Education and Training in Analysis, Synthesis and Repackaging of Information*. Paris: UNESCO, 1986.

——. "Processes and Problems in Information Consolidation". *Information Processing and Management*, Volume 22 (1) 1986: 45-60.

Saracevic, T. and Wood, J.B. *Consolidation of Information: A Handbook on Evaluation, Restructuring and Repackaging of Scientific and Technical Information*. Paris: UNESCO, 1981.

Seetharama, S. *Guidelines for the Establishment of Information Consolidation Units*. Paris: UNESCO, 1984.

——. *Information Consolidation and Repackaging: Framework, Methodology, Planning*. New Delhi: Ess Ess Publications, 1997.

Weisman, Herman M. *Information Systems, Services and Centres*. New York: Becker and Hayes, 1972.

Bibliography

Agada, John. "Information Technology and the Preparation of Liberians for the Information Society: Questions for Developing Countries". *Personnel Training and Education*, Vol. 7 (3) 1990.

Arora, Renu. "Technical Writing for Library and Information Professionals". In: *Challenges and Changes in Librarianship: Papers in honour of Professor Sewa Singh*. Edited by A. Amudhavalli and Jasmer Singh. New Delhi: B.R. Publishing Corporation, 2010. 2 Volumes.

Ashworth, W. "Abstracting". In: *Handbook of Special Librarianship and Information*. 4th ed. London: ASLIB, 1975.

Atherton, Pauline. *Handbook for Information Systems and Services*. Paris: UNESCO, 1977.

Bearman, T.C. "Educating the Future Information Professional". *Library Hitech News*, Vol. 18, 1984: 27-40.

Bhat, Ishwara M. "Marketing of Library and Information Services at British Council Library Network in India". *DESIDOC Bulletin of Information Technology*, Vol. 18 (3) May 1998: 29-33.

Bhattacharyya, G. "Information Analysis for Consolidation: Some Basic Considerations". *DRTC Annual Seminar*. Volume 18. 1981, Paper A.

——. "Knowledge and Skills for Secondary Information Work and Service: Foundation of Curricula and Syllabi". In:

Restructuring of Library and Information Science Curriculum. Edited by A. Tejomurty. New Delhi: Concept Publishing Company, 1994.

Borko, H. and Chatman, S. "Criteria for Acceptable Abstracts: A Survey of Abstractors". *American Documentation*, Vol. 14, 1963: 149-60.

Brandreth, M. Specialized Information Analysis Centres in International Development. Report of a Meeting held at Montebellow, Quebec, Canada, 4-8 October, 1982. IDRC Manuscript Series. March 1984.

Bryson, Jo. *Effective Library and Information Centre Management*. Bombay: Jaico Publishing House, 1996.

Buckland, Michael K. Redesigning Library Services: A Manifesto. Chicago: American Library Association, 1992.

Bunch, A. The Basics of Information Work. London: Clive Bingley, 1984.

Campbell, W.G. and Ballou, S.V. *Form and Style*. Boston: Houghton Miffin, 1978.

Carbo, Toni. "The Future of Liberianship: A View from a School of Library and Information Science". *Library Connect Newsletter*, Vol. 1 (4) December 2003: 5-6.

Cronin, Blaise. *The Marketing of Library and Information Services*. London: ASLIB, 1981.

Devarajan, G. *Library Information User and Use Studies*. New Delhi: Beacon Books, 1995.

Devarajan, G. *Technical Communication for Information Managers*. New Delhi: Ess Ess Publications, 2012.

Documentation Research and Training Centre, Bangalore. *Information Analysis and Consolidation: Principles, Procedures and Products*, by G. Bhattacharyya and M. A. Gopinath. Annual Seminar. 18. Bangalore: DRTC, 1981.

Gerson, Sharon J. and Gerson, Steven M. *Technical Writing: Process and Product*. 6th ed. New York: Prentice-Hall, 2007.

Goi, Chai Lee. "A Review of Marketing Mix: 4Ps or More"? *International Journal of Marketing Studies*, Vol. 1 (1) May 2009: 2-15.

Goode, W.J. and Hatt, P.K. *Methods in Social Research*. New York: McGraw-Hill, 1952.

Greer, R.C., Agada, J. and Grover, R. "Staffing: A Model for Library and Information Agencies". *Library Administration and Management*, Vol. 8 (1) 1994: 35-42.

Guha, B. *Documentation and Information Services, Techniques and Systems*. 2nd rev. ed. Calcutta: World Press, 1983. Reprinted 1999.

Gupta, D.K. and Jambekar, A. "Developing a Customer-Focus Approach to Marketing of Library and Information Services". *DESIDOC Bulletin of Information Technology*, Vol. 22 (3) May 2002: 5-14.

International Federation of Library Associations and Institutions (IFLA). *Guidelines for Professional Library/Information Education Programs*. Latest revision 2009. (Retrieved from http://www.ifla.org/VII/s23/bulletin/guidelines.htm on 06-06-2013).

Iwhiwhu, Enemute Basil. "Information Repackaging and Library Services: A Challenge to Information Professionals in Nigeria". *Library Philosophy and Practice-electronic journal*, 2008.

Jain, Abhinandan, et al. *Marketing Information Products and Services: A Primer for Librarians and Information Professionals*. New Delhi: Tata McGraw-Hill, 1999.

Janakiraman, M. "Marketing Mix in the Context of Library and Information Products and Services". *DESIDOC Bulletin of Information Technology*, Vol. 18 (3) May 1998: 11-19.

Katz, William A. *Introduction to Reference Work*. 8th ed. New York: McGraw-Hill, 2001. 2 Volumes.

Kawatra, P.S., Ed. *World Encyclopedia of Library and Information Science Education.* New Delhi: Crest Publishing House, 1994. 3 Volumes.

Kertesz, F. "Collaboration Between IACs at a Large Multipurpose Laboratory". Quoted in Weisman, Herman, H. *Information Systems, Services and Centres*, New York: Becker and Hayes, 1972.

——. *Guidelines for Establishing and Operating Information Analysis Centres*, Paris: UNESCO, 1978. Prentice-Hall of India, 2007.

Kotler, Philip. *Marketing Management Analysis, Planning, Implementation and Control.* New York: Prentice-Hall, 1997.

Kotler, Philip and Keller, Kevin L. *Marketing Management.* 12th ed. New Delhi: Prentice-Hall of India, 2007.

Kumar, P.S.G. *Information Analysis, Repackaging, Consolidation and Information Retrieval.* New Delhi: B.R. Publishing Corporation, 2004.

Kunz, Werner, et al. *Methods of Analysis and Evaluation of Information Needs.* Munchen: Verlag Documentation, 1977.

Lancaster, Geoff and Withey, Frank. *Marketing Fundamentals.* New York: Elsevier Science and Technology, 1994.

Lancaster, F.W. *Guidelines for the Evaluation of Information Systems and Services.* Paris: UNESCO, 1978.

——. *Guidelines for the Preparation of Training Courses, Workshops and Seminars*, Paris: UNESCO, Bureau of Studies and Programming, 1983.

——. *If You Want to Evaluate Your Library.* 2nd ed. Champaign, IL: University of Illinois, 1993.

——. *Information Retrieval Systems: Characteristics, Testing and Evaluation.* 2nd ed. New York: John Wiley, 1979.

——. *The Measurement and Evaluation of Library Services.* 2nd ed. Arlington, Va: Information Resource Press, 1993.

Landau, Herbert B., et al. "An Information Transfer Model to Define Information User and Outputs with Specific Application to Environmental Technology". *Journal of the American Society for Information Science*, Vol. 33 (2) March 1982: 82-91.

Laxman Rao, N. et al. eds. *Changing Dimensions* of LIS Education: Seminar Papers. XVIII IATLIS National Seminar, Andhra University, Visakhapatnam, 12-14 October, 2001. Hyderabad: IATLIS, 2001.

Lea, Peter W. and Day, Alan, eds. *The Reference Sources Handbook*. 4th ed. London: Library Association Publishing, 1996.

Naylor, R.J. "Core Competencies: What They are and How to Use Them". Public Libraries, Vol. 39, 2000: 108-14.

Neelameghan, A. *Presentation of Ideas in Technical Writing*. New Delhi: Vikas Publishing House, 1975.

Neelameghan, A. and Seetharama, S. "Reference Framework: An Aid to Information Analysis and Consolidation". In: *DRTC Annual Seminar*. 18, 1981, Paper BD.

Parekh, Harsha. *Bridging the Divide: Perspectives of an Information Professional*. Compiled by Bharati Sen and Nila Jani. Mumbai: SHPT School of Library Science, 2003.

Ranganathan, S.R. *Reference Service*. 2nd ed. Bombay: Asia Publishing House, 1961.

Resnik, Lynda. Rubies in the Orchard: How to Uncover the Hidden Gems in Your Business. Crown Publishing, 2009.

Rowley, Jennifer. *The Basics of Information Systems*. 2nd ed. London: Library Association Publishing, 1996.

——. *The Electronic Library*. 4th ed. London: Library Association Publishing, 1998.

Saracevic, T. *A Course in Information Consolidation: A Handbook for Education and Training in Analysis, Synthesis and Repackaging of Information* (Preliminary Version). Paris: UNESCO, 1986.

——. "Processes and Problems in Information Consolidation". *Information Processing & Management*, Vol. 22 (1) 1986: 45-60.

Saracevic, T. and Wood, J.B. *Consolidation of Information: A Handbook on Evaluation, Restructuring and Repackaging of Scientific and Technical Information*. Paris: UNESCO, 1981.

Seetharama, S. "Guidelines for Abstracting". *Eighth IASLIC Seminar*, held at Madras, 1978. Calcutta: IASLIC, 1978.

——. *Guidelines for the Establishment of Information Consolidation Units*. Paris: UNESCO, 1984.

——. *Information Consolidation and Repackaging: Framework, Methodology, Planning*. New Delhi, 1997.

Shera, J.H. *Foundations of Education for Librarianship*. New York: Becker and Hayes, 1972.

Singh, Sewa. *Challenges and Changes in Librarianship....* Edited by A. Amudhavalli and Jasmer Singh. New Delhi: B.R. Publishing Corporation, 2010. 2 Volumes.

——. "Systems Analysis: Concept Design, Model Tools and Techniques". In: *Knowledge Management in Digital Era*. Edited by A. Chandel. New Delhi: Westville Publishing House, 2011, pp. 144-76.

Special Libraries Association. *Competencies for Special Librarians of the 21st Century*: Full report, Submitted by Joanne Marshall, et al. Washington: SLA, 1996. Revised June 2003.

Special Libraries Association. *Competencies for Special Librarians of the 21st Century: Library and Information Studies Programs Survey: Final Report*. Washington: SLA, 1998.

Steiner, G.A. *Top Management Planning*. Canada: Collier Macmillan, 1969.

Struges, P. and Chimsen G. "Information Repackaging in Malawi". *African Journal of Library, Archives and Information Science*, Vol. 6 (2) 1996: 85-93.

Subramanyam, K. "Review Literature". In: *Encyclopedia of Library and Information Science*. New York: Marcel Dekker, 1979. Vol. 26, pp. 461-74.

Toffler, Alvin. *The Third Wave*. New York: Morrow, Williams & Company, 1980.

Valls, J. *Information Services for Developing Countries*. Bangakok: Asian Institute of Technology, 1983.

——. "Some Aspects of Information Consolidation in and for Developing Countries". Paper Presented at the Third Meeting of UNISIST Working Group on Information Analysis and Consolidation. Kuala Lumpur, 12-16 September, 1983.

Vickery, B.C. "Concepts of Documentation". *Journal of Documentation*, Vol. 34, 1978: 279-87.

Webb, S.P. *Creating an Information Service*. London: ASLIB, 1985.

Weisman, Herman M. *The Importance of Information Analysis Centres in the Performance of Information Services*. Washington, D.C.: National Institute of Education, 1973.

Weisman, Herman M. *Information Systems, Services and Centres*. New York: Becker and Hayes, 1972.

Weiner, N. *Cybernetics*. New York: John Wiley, 1948.

Wetherbe, J.C. Systems Analysis and Design. 3rd ed. St Paul, USA: West Publishing Company, 1988.

Whittaker, Kenneth. *The Basics of Library Based User Services*. London: Library Association, 1993.

Woodward, A.M. "The Roles of Reviews in Information Transfer". *Journal of the American Society For Information Science,* Vol. 28 (3) May 1977: 175-80.

Index

F

G

H

I